P9-DMU-357

THE CONSCIOUSNESS INSTINCT

THE CONSCIOUSNESS INSTINCT

UNRAVELING

THE

MYSTERY

OF HOW

THE BRAIN

MAKES

THE MIND

Michael S. Gazzaniga

FARRAR, STRAUS AND GIROUX | NEW YORK

Farrar, Straus and Giroux
175 Varick Street, New York 10014

Library of Congress Cataloging-in-Publication Data
Names: Gazzaniga, Michael S., author.
Title: The consciousness instinct : unraveling the mystery of how the brain
    makes the mind / Michael S. Gazzaniga.
Description: New York : Farrar, Straus and Giroux, 2018. | Includes
    bibliographical references and index.
Identifiers: LCCN 2017038333 | ISBN 9780374715502 (cloth) |
    ISBN 9780374128760 (e-book)
Subjects: LCSH: Brain. | Mind and body. | Consciousness. | Cognitive
    neuroscience.
Classification: LCC QP376 .G386 2018 | DDC 612.8/2—dc23
LC record available at https://lccn.loc.gov/2017038333

Designed by Richard Oriolo

Our books may be purchased in bulk for promotional, educational, or
business use. Please contact your local bookseller or the Macmillan
Corporate and Premium Sales Department at 1-800-221-7945, extension
5442, or by e-mail at MacmillanSpecialMarkets@macmillan.com.

www.fsgbooks.com
www.twitter.com/fsgbooks • www.facebook.com/fsgbooks

1 3 5 7 9 10 8 6 4 2

For Leonardo,
consciousness unfolding if I ever saw it

CONTENTS

THE CONSCIOUSNESS INSTINCT

INTRODUCTION

MAGINE, IF YOU CAN, being conscious of only one moment—right now. This moment exists without a past or a future. Now imagine life being a series of these moments, each existing in some kind of isolation from all other moments, not connected by subjective time. Imagine being temporarily frozen in each of the moments that together make up normal living. It is hard to imagine this scenario because our minds travel back and forth through time so fluidly, like a ballerina in *The Nutcracker*. One moment serves as the grist for the next planned action, which is in turn weighed in the present against our past experience. It is hard to imagine this ever not being true. And yet, conk your head in the right way and that might be you, still able to understand the idea of having a past and future, but unable to place yourself in your own past or future. Weird, if true. No past, no future, only the present.

In this book I will take you on a journey through a world where hard-to-imagine alterations in what we call conscious experience are commonplace. The neurologic ward of every hospital is replete with disruptions of normal conscious experience. Each of these cases tells us something about how our brains are organized to deliver up our cherished consciousness, moment by moment. Each example of disruption cries out to be understood, to be used to deduce a coherent story about how our brains build and produce the everyday joys of being conscious. In the past, folks were content to tell stories of these bizarre phenomena. Here in the twenty-first century, it is not enough to simply describe the basketful of intriguing disorders. In this book, my goal is to move forward on the problem of consciousness, and I will try to illuminate how our exquisitely evolved brain is organized to do its magic. In short, I want to examine how matter makes minds.

A few years back, a work trip found me passing through customs at Heathrow in London. The passport control officer, an attentive British civil servant, dutifully asked me my name and business and reason for coming to the United Kingdom. I told him I did brain research and was on my way to Oxford for a meeting. He asked if I was aware of the two different functions of the cerebral hemispheres, the left and the right. I somewhat proudly said that not only had I heard about it, I was in part responsible for the work. As he perused my passport, he asked what the meeting in Oxford was going to be about. I responded, with an air of authority, "About consciousness."

The agent closed the passport and, handing it back, asked, "Have you ever thought about quitting while you are ahead?"

Apparently not. Some of us are naturally incautious in our desire to wonder about our nature. Working away in the mind/brain sciences as I have for sixty years makes me painfully aware that we humans have not yet grasped the problem in its fullness. Still, it is in our nature to think about who and what we are and what it means to be conscious. Once bitten by the question, we spend our lives gnawed by the desire for an answer. Yet, when we try to grab hold of the problem of consciousness, it seems to dissolve like fog. Why has the quest to understand consciousness been so difficult? Do the lingering ideas of the past block us from seeing clearly how it comes about? Is consciousness just what brains do? Just as a pocket watch with all of its gears tells us the time, do brains with all their neurons just give us consciousness? The history of the topic is vast, swept by pendulum swings between the pure mechanists and the hopeful mentalists. Surprisingly, twenty-five hundred years of human history have not resolved the question or taught our species how to frame an understanding of our personal conscious experience. Indeed, our core ideas have not changed that much. While thinking explicitly about consciousness was ignited by Descartes three hundred years ago, two overarching and contradictory notions—that the mind either is part of the brain's workings or works somehow independently of the brain—have been around seemingly forever. Indeed, these ideas are still with us.

In recent years the topic of consciousness has become red-hot once again. At the same time, and despite the modern avalanche of new data, there are few, if any, generally accepted proposals on how the brain builds a mind and, with it, conscious experience. The goal of this book is both to break free from this quagmire and to present a new view of how to conceptualize consciousness. Our journey includes knowledge gained not only from neurology but also from evolutionary and theoretical biology, engineering and physics, and, of course, psychology and philosophy. Nobody said the search was going to be easy. But the goal—understanding how nature pulls off the trick of turning neurons into minds—is attainable. So hold on to your hat!

Plainly stated, I believe consciousness is an instinct. Many organisms, not just humans, come with it, ready-made. That is what instincts are, something organisms come with. Living things have an organization that allows life and ultimately consciousness to exist, even though they are made from the same materials as the non-living natural world that surrounds them. And instincts envelop organisms from bacteria to humans. Survival, sex, resilience, and walking are commonly thought to be instincts, but so, too, are more complex capacities such as language and sociality—all are instincts. The list is long, and we humans seem to have more instincts than other creatures. Yet there is something special about the consciousness instinct. It is no ordinary instinct. In fact, it seems so extraordinary that many think only we humans can lay claim to it. Even if that's not the case, we want to know more about it. And because we all have it, we all think we have insight into it. As we will see, it is a slippery, complex instinct situated in the universe's most impenetrable organ, the brain.

The word "apple" is a noun; it signifies a real, physical object. The word "democracy" is a noun as well, and it describes something a little harder to pin down, a state of societal relationships. It is easy for me to show you an apple, its physical reality. It is hard for me to show you the physical reality of a democracy. How about "instinct," another noun? All three are definable *somethings*, whether they are objects or concepts, managed by the brain. We have lots of these somethings, but where are they in the brain? Are some best represented as actual structures in the brain and others best represented by the processing actions of structures? Indeed, what is the physical reality of an instinct? Is it tangible, like an apple, or elusive, like a democracy?

Complex instincts are more like democracies; they are identifiable but not easily localized. They emerge from the interaction of simple instincts but are not those things themselves—just as an intricate pocket watch chugs away at keeping time, yet time itself is impossible to find in the watch. In order to understand how the watch relates to time, you have to describe its principles of design, its architecture, not just list all its springs and gears. The same is true of the consciousness instinct. Don't think that if consciousness is an instinct, there will be a single, unitary, discrete brain network generating that phenomenal self-aware state we all relish. It is not like that at all. When we visit the neurologic wards, armed with our ideas, you'll recognize right away that patients who suffer from dementia, even severe dementias, are conscious. These patients with widely distributed brain lesions, a level of disruption vast enough to bring any computing machine to its knees, remain conscious. In one hospital room after another, each harboring a patient with a focal or a diffuse brain impairment, consciousness purrs along. After a tour of the wards, it begins to look like consciousness is not a system property at all. It is a property of local brain circuits.

In this book's first section, we will see how nature became an "it," a thing separate from us that can be studied and understood in objective terms. We trace this idea all the way through Descartes up to modern times and the dawn of modern biology. Surprisingly, most modern scientific thinking has looped back to build on the ideas of the ancient Greeks, and holds essentially the same models, which link the mental and physical inexorably together in one system. Modern science has started to pursue the same goal the Greeks sought, but so far it, too, has fallen short. Again, new ideas are needed, and this book takes a shot.

In Part II, some modern principles of brain functioning are introduced that I feel should guide our journey into how neurons produce minds. It is amazing to me how the "brain as a machine" metaphor, first proposed by Descartes and wholly adopted by most of modern science, has led us to believe that the entire machine is needed to perform many of its functions. In fact, we are each a confederation of rather independent modules, orchestrated to work together. To understand how those modules collaborate, we need to know about the overall architecture of the system, an architecture called "layering" that will be familiar to many readers, such as computer scientists. And finally, we'll pay that visit to the neurologic clinic to test this formulation. There we will discover that our modular brain with its layered architecture is managing our consciousness from . . . everywhere in its local tissues, over and over again. There is not one centralized system working to produce the grand magic of conscious experience. It is everywhere, and you can't seem to stamp it out, not even with a wide-ranging brain disease like Alzheimer's.

In Part III, I confront that nagging issue at the core of this mind/brain business: How do neurons gin up mind? How do those squishy bundles of wet tissue make you and me mental? It turns out there are gaps throughout our understanding of the physical world. We study one level of organization and then another, but in fact we don't understand how the two different levels work together. There is a notorious gap between life and inanimate matter, between mind and brain, between the quantum world and our everyday world. How can those gaps be closed? It looks to me like physics can help us.

Finally, I offer a perspective on how the modules, layers, and gaps play out to yield what we call conscious experience. The psychology professor Richard Aslin once commented to me that he felt the idea of "consciousness" was a proxy for a whole host of variables correlated with our mental lives. We use "consciousness" as shorthand to easily describe the functions of a multitude of inborn, instinctual mechanisms such as language, perception, and emotion. It becomes evident that consciousness is best understood as a complex instinct as well. All of us come with a bucketful of instincts. Our incessant thought pattern jumps around. We have feelings about one idea, then its opposite, then our family, then an itch, then a favorite tune, then the upcoming meeting, then the grocery list, then the irritating colleague, then the Red Sox, then . . . It goes on and on until we learn, almost against our natural being, to have a linear thought.

Conscious linear thinking is hard work. I'm sweating it right now. It is as if our mind is a bubbling pot of water. Which bubble will make it up to the top at any given moment is hard to predict. The top bubble ultimately bursts into an idea, only to be replaced by more bubbles. The surface is forever energized with activity, endless activity, until the bubbles go to sleep. The arrow of time stitches it all together as each bubble comes up for its moment. Consider that maybe, just maybe, consciousness can be understood only as the brain's bubbles, each with its own hardware to close the gap, getting its moment. If that sounds obscure, read the book to find out for yourself whether you can see it this way, too. Importantly, enjoy your thoughts as they bubble up to the surface of your own consciousness.

PART I:

GETTING READY FOR

MODERN THOUGHT

1.

HISTORY'S RIGID, ROCKY, AND GOOFY WAY OF THINKING ABOUT CONSCIOUSNESS

"Speak English!" said the Eaglet. "I don't know the
meaning of half those long words, and, what's
more, I don't believe you do either!"
—Lewis Carroll, *Alice's Adventures in Wonderland*

SIGMUND FREUD DIED the year I was born—1939. That year there were
a lot of zany ideas being kicked around about the nature of our psychological lives, many of them dreamed up by Freud himself. He is not
popularly remembered as such, but Freud was a biologist at heart, a reductionist. He was committed to the belief that the brain generated the mind in
a deterministic way, a view shared by many of today's neuroscientists.
Now we recognize that many of his ideas were pure fantasy, but up until the
1950s they were so broadly accepted that they were the dominant testi-
mony for psychological issues in a U.S. court of law!

It has been in my lifetime, not Freud's, that humankind has learned the most about how the brain does its tricks. Wild speculation about the forces governing our mental lives has given way to specific knowledge about the molecular, cellular, and environmental influences that underlie our existence. Indeed, the past seventy-five years of research have provided a wealth of information about the brain, sometimes even yielding organizing principles. I am sure Freud would have reveled in our new world and would have gladly put his incredible imagination to work on the new science of the brain. Yet the deep puzzles that faced scientists of all stripes in the previous century, and indeed going back to the ancient Greeks, are still present today. How on earth does lifeless matter become the building blocks for living things? How do neurons turn into minds? What should be the vocabulary used to describe the interactions between the brain and its mind? When humankind finds some answers, will we be disheartened by what they are? Will our future understanding of "consciousness" simply not be fulfilling? Will it be simple yet cold and harsh?

Wading into the history of the study of consciousness is daunting. For one thing, it is littered with the complex and abstract writings of philosophers. Even John Searle, one of today's leading philosophers of consciousness, has admitted: "I probably should read more philosophy than I do. But I think a lot of works of philosophy are like root-canal work, you just think you've got to get through that damn thing."[1] Add to that the view of the great philosopher David Hume, who provided strong arguments that most of the questions asked by philosophers simply couldn't be answered using the methodologies of logic, mathematics, and pure reason. Nonetheless, philosophers got us thinking about the mind, the soul, and consciousness. From ancient times on, they have had a huge influence.

"Consciousness" is a relatively modern idea. The very word, as now broadly used in dozens of contexts (Marvin Minsky would call it a "suitcase word" because it is packed full of various meanings), was invented in its modern sense only in the mid-seventeenth century by René Descartes. It does have origins in the Greek word *oida*—"to have seen or perceived and hence to know"—and the Latin equivalent *scio*, "to know." But the ancients did not have an explicit concept of consciousness. There was interest in how the mind worked, where thoughts came from, and even whether a purely physical process was involved, but most early thought wound up concluding that mental life was the product of an immaterial spirit. And when consciousness is framed as immaterial spirit, it's hard to start thinking about underlying mechanisms!

Over the centuries, the concept of the mind and the concept of the soul have been involved in an on-again, off-again relationship. For most of written history, the very idea that personal psychological reality was a *thing*, a something to be studied, was largely nonexistent. Our brains, our thought structures, and our emotions presumably haven't changed, so what were we humans thinking about? But, as will become evident, the concept of consciousness has radically changed over the past twenty-five hundred years. Its ethereal beginnings and its current meaning have little to do with each other.

We humans need a new way to think about the problem, and with luck, this book may offer some new beginnings. First, however, as is always the case, it's best to look back before plunging forward.

Early Stirrings: Successes and Blunders

The ancient Egyptians and Mesopotamians were the Western world's philosophical forebears. In their concept of the world, nature was not an opponent in life's struggles. Rather, man and nature were in the same boat, companions in the same story. Man thought of the natural world in the same terms as he thought of himself and other men. The natural world had thoughts, desires, and emotions, just like humans. Thus, the realms of man and nature were indistinguishable and did not have to be understood in cognitively different ways. Natural phenomena were imagined in the same terms as human experience: generous or not so much, dependable or spiteful, and so on. These ancients of the Near East did recognize the relation of cause and effect, but when speculating about it they came from a "who" rather than a "what" perspective. When the Nile rose, it was because the river wanted to, not because it had rained. There was no science to suggest otherwise.

Not so with the ancient Greeks. The earliest Greek philosophers were not priests charged by their communities to consider spiritual matters, as they were in the Near East. They were not professional seers. They were a bunch of amateurs puttering around in their garages unconstrained by dogma, curious about the natural world, and happy to share their thoughts. When they started to ask about their origins, they did not ask "who" the progenitor was, they asked "what" the first cause was. This was a monumental change of viewpoint for humankind that the archaeologist and Egyptologist Henri Frankfort called "breathtaking":

> [T]hese men proceeded, with preposterous boldness, on an entirely unproved assumption. They held that the universe is an intelligible whole. In other words, they presumed that a single order underlies the chaos of our perceptions and, furthermore, that we are able to comprehend that order.[2]

Frankfort goes on to explain how the Greek philosophers were able to make this leap: "The fundamental difference between the attitudes of modern and ancient man as regards the surrounding world is this: for modern, scientific man the phenomenal world is primarily an 'It'; for ancient—and also for primitive—man it is a 'Thou.'"

A "Thou" is a someone with beliefs, thoughts, and desires, doing their thing, not necessarily stable or predictable. On the other hand, "It" is an object, not a friend. "It" can be related to other objects in whatever seems the most reasonable organization. One can build and expand on these relationships and seek universal laws that govern behavior and events under predictable, prescribed conditions. Seeking the identity of an object is an active process. On the contrary, understanding a "Thou" is a passive process in which one first receives an emotionally charged impression. A

"Thou" is unique and unpredictable and known only insofar as it reveals itself. Each "Thou" experience is individual. You can coax a story or a myth from an interaction with a "Thou," but you cannot draw a hypothesis. The transition away from "Thou" and toward "It" made scientific thinking possible.

The Greeks' huge advance in perspective created an atmosphere that catapulted Aristotle into a scientific life. Aristotle's stance was that the job of science was to account objectively for the "why" of things, which led to his doctrine of causality. For him, scientific knowledge about something (say, some X) included all the ways the "why" question could be answered: if X was caused by Y, or if Y was at least a necessary condition in order for X to happen, then this is the type of assertion that belongs to science. He postulated four causal categories: material, formal, efficient, and final. So if one were to ask "Aristotle, why a cart?," he would tell you the material cause was wood, the formal cause was its blueprint, the efficient cause was its construction, and its final cause was . . . he just wanted one.

For Aristotle, the natural world was a web of what biological theorist Robert Rosen calls *causal entailments*: X comes with all its Ys. Rosen points out that Aristotle's whole idea was to show that no one mode of explanation sufficed to understand anything, because the causal categories do not entail each other. For example, knowing how to build something does not entail understanding how it works; knowing how something works does not entail knowing how to build it. Also, for Aristotle, science was content-determined. It was independent of the method by which it was studied.

The scientific method as practiced today is a formal system in which a hypothesis produces its inferences, that is, its effects: the hypothesis entails its effects. Another way to say this is that the cause comes before the effect. This presents a problem when asking Aristotle's final-causation "why" question. Let's go back to "Why the cart, Aristotle?" Why did Aristotle have a cart parked in front of his home when hours earlier it had been parked at Acropolis Depot? He had seen the cart (which entailed the effects of the material, formal, and efficient causes) and wanted it. Here, the tables were turned and the effect came before the cause. This is a no-no in the Newtonian world, where a state can only entail subsequent states. Thus, Aristotle's final causation, as a separate category, was lost to science. We will see later what harm this has done to biology.

Among other things, Aristotle wanted to know more about the human body and how it worked. This was a bit challenging, since the Greeks had a taboo against human dissection. Aristotle skirted this issue by performing numerous animal dissections. From what he learned, he devised a system of classifying organisms, the *scala naturae*, a graded hierarchical scale based on the type of "soul" each possessed. At the base were plants, which he posited have a *vegetative soul* responsible for growth and reproduction. Needless to say, man sits at the top of the *scala naturae*.

Aristotle didn't stop there. He proposed that animals possess a *sensitive soul* powering self-movement, perception, sensation, appetite, and emotion. Unique to humans and nested within the sensitive soul is a *rational soul* that provides us with the special powers of reason, rational will, thought, and reflection and sets us apart from those lower on the *scala*. Most important, and reflecting the revolution in human thinking, the "knowledge" of these powers arrived at not by sheer introspection or mental meanderings, but by observing how one connects with the surrounding world. The "it," that is, an object such as the world around us, could be studied and examined. We forget that this very idea, now commonly accepted, didn't exist a few thousand years ago! Clearly, ideas do have consequences, and, happily, we continue to be captivated by the idea and power of scientific observation.

Aristotle got the process of science right, but his conclusions about where thoughts come from were all off. If a modern student had made a mistake like the one he made, the student would have failed the course. Aristotle knew from the actions of animals and humans that they can perceive the world. From his dissections, Aristotle noted that some animals had no visible brains at all. He concluded, therefore, that the brain appears to be of not much account. The first thing he saw appearing in the embryos he studied was the heart, so he put the soul there, which in the case of humans included the rational soul. Aristotle did not mean "soul" in a spiritual sense, as he did not think it continued on after death. He meant the organ that gives rise to sensation, to our knowledge of the world. He thought that the rational soul, which was the source of human intellect, required some perceptual mechanisms; therefore, it required a body with its parts and organs. Yet he did not think that there was a body part or an organ that thinks. Aristotle never even mumbled the word "conscious," but he did ask, "How do we know our own perceptions?" Overall, Aristotle got the ball rolling and got people thinking about humankind's physical nature.

The monumental stirrings that started in Greece were quickly exported. In 322 B.C., not long after Aristotle died, Herophilus and Erasistratus, two Greek physicians living in Alexandria, defied the taboo on dissecting human bodies and went at it. They became the first to discover the nervous system and write about it. They also found the ventricles, the empty chambers inside the brain. Herophilus decided that these chambers must be where the intellect was located, and that from them, spirits flowed down through hollow nerves out to the muscles, making them move. While they didn't get it exactly right, they are commonly credited with being the first neuroscientists. Unbelievable as it now may seem, the Greek culture that engineered and built the Parthenon didn't know about brains. And the Egyptian culture that engineered and built the pyramids didn't know how the brain worked at all.

History rattled along for another four hundred years, a microsecond in evolutionary time. Rome became the dominant force in the Mediterranean and somehow was able to attract the wondrous physician Claudius Galenus (Galen) from Pergamum, a Greek city on the Aegean coast of modern-day Turkey. Galen finished his medical training an empiricist, having immersed himself in the teachings of Herophilus and Erasistratus in Alexandria, now under Roman rule. In ancient Greece, the Empiric school of medical practice relied on the observation of phenomena and on experience, not on dogmatic dicta. Galen returned to Pergamum for his first job: gladiator doctor. Because the Romans, like the Greeks, did not allow human dissection, Galen never did any. Instead, he honed his knowledge of anatomy and surgery with the gory remains of his patients and with daily animal dissections, primarily on Barbary macaques. He took his firsthand knowledge; a healthy helping of the teachings of his distant mentors, Herophilus and Erasistratus; and a pinch of Hippocrates' theory that the body was composed of four humors, and combined them into a new conception of the body and its machinations. He earned himself a stellar reputation. Soon he was on his way to Rome, and his growing fame led him to become the personal physician to the emperor, Marcus Aurelius.

Galen's contributions to medicine are stunning. He was the first to recognize that there is a difference between arterial and venous blood. We now know that arterial blood is rich with oxygen, while venous blood carries much less (your tissues have stolen it so they can breathe), a difference that is exploited in the functional magnetic resonance imaging (fMRI) studies of the brain that are the cornerstone of modern neuroscience. Galen gave the first depiction of the four-chambered human heart; he updated the knowledge of the circulatory system, the respiratory system, and the nervous system. He made some anatomical blunders, of course—one being a meshwork of blood vessels, the *rete mirabile*, which he located at the base of the human skull, based on dissections of oxen. This was a major mistake and a cautionary tale about inductive reasoning. As was shown years later, humans flat-out don't have a *rete mirabile*!

Nonetheless, Galen understood that food and breath are necessary for human life, and maintained that the body transforms them into the flesh and spirit. Amalgamating the works of Hippocrates, Plato, Socrates, and Aristotle, Galen came up with the idea of a material tripartite soul. Using Plato's designations of the rational, spiritual, and appetitive souls, he assigned each one an anatomical location: the rational soul was in the brain, the spiritual soul was in the heart, and the appetitive soul was in the liver. Each performed a separate function. The appetitive soul controlled the natural urges of the body, such as hunger and thirst, survival instincts, and bodily pleasures. It was animated by natural spirits. The spiritual soul contained the emotions and passions and was animated by a vital spirit that somehow formed in the heart from blood and air delivered via the lungs. The rational soul controlled cognition such as perceptions, memory, decision making, thought, and voluntary action. Galen saw no distinction between the mental and the physical. One can begin to see the groundwork being laid for such modern ideas as conscious versus subconscious, the id and the ego, the rational and the intuitive. The specifics are different, but the underlying ideas were emerging even in A.D. 200.

Galen took a stab at mechanism. He envisioned a vital spirit, a life-giving force that enters the body and is purified in the *rete mirabile*. The purified spirit then flows into the ventricles of the brain, where it becomes an animal spirit and enables the rational soul's cognition. While Galen got the organ for cognitive functions right, he did not really understand it. He located all the processing in the empty ventricles. That is like saying the best part of the doughnut is the hole.

Still, one of Galen's major contributions to the future of medicine was the notion that different organs perform different functions. Beginning to differentiate the body's organs into various machines performing separate functions was a tremendous idea. Today, one of the goals of modern neuroscience is to discover what functions the various parts of the brain perform. With each century, neuroscience continues to get more and more specific about which particular brain systems contribute to our overall mental life. In true reductionist style, Galen did not distinguish between the physical and the mental, yet at the same time he held on to the idea of an immortal soul. Time and time again, as we will see, the brilliant forebears of modern neuroscience abandoned their fierce reasoning skills and, deus ex machina, threw in a spook at the end of their analysis.

Throughout his life, Galen was a firm believer in personal observation and experimentation over established teachings, but he didn't completely practice what he preached. His epistemology was rooted in his philosophical training, which included the teaching of Plato, Aristotle, and the Stoics, and he mixed and matched some of it with his observations to create an overarching theory of medicine. Yet he most likely would have been completely dismayed by the influence he had on medicine for the next thirteen hundred years. Galen's findings were taken as gospel for over a millennium! Some of his ideas were instituted as doctrines by the new Christian church. In the Old Testament, the soul died with the body, just as Aristotle had asserted. The new Christians, however, had a different view of the soul. They conceived of it as immortal, living beyond the life of the body, just as Plato and Socrates had suggested. Although Galen believed there to be no distinction between the mental and the physical, the Christians liked Galen's idea that the soul was located in the airy ventricles, tucked away from the lusty, sinning body. So that became the Church's doctrine of the bodily location of the newly immortal and immaterial soul. Sensation was ensconced in the front ventricle, understanding was in the middle ventricle, and memory brought up the rear.

From the early Greeks through Galen's period of influence, a period of seventeen centuries of human thought, thinking about the nature of human existence found us wallowing in high-powered confusion. Most of the talk was about souls, not minds—and certainly not consciousness. Plato and Socrates argued for a tripartite immortal soul, partly rational, partly spiritual, and partly appetitive. Aristotle also reasoned we had souls, but he said they were not immortal. The early students of the brain, and of anatomy in general, went back to saying that they were immortal but that there was no difference between the mental and the physical. Ideas die hard, even in the light of an emerging science. As we shall see, these primitive ideas are still in play today.

Setting the Stage for Descartes and the Idea of Mind/Body Dualism

It wasn't until the sixteenth century that Galen's anatomy was challenged by a young anatomist, Andreas Vesalius, based at the University of Padua. Vesalius started scratching his head when he compared his own human dissections with Galen's drawings. Luckily for him (and modern science), he suffered no human-dissection taboos, and the local judge had no qualms about sending him the cadavers of condemned criminals. Vesalius came to the realization that not only had Galen never dissected a human, but much of his anatomy was just plain wrong. Vesalius did not have the greatest tools when it came to dissecting the brain. He sawed it in slices from the top down, mauling the lower sections as he went, somewhat like slicing a ball of mozzarella di buffala with a dull knife. But one thing became perfectly clear: there was no *rete mirabile*. One aspect of science that we have learned over the course of centuries that is hugely important is to check and double-check an earlier claim.

A few years earlier another anatomist, Niccolò Massa at the University of Bologna, had discovered that the ventricles were filled not with airy spirits but with fluid. Now Vesalius found that they were not the perfect spheres with fleshy vaults that Galen had described. Enough things were wrong with Galen's descriptions that Vesalius had to rewrite (or redraw) the book, so to speak. With the help of apprentices from Titian's workshop in Venice, *De Humani Corporis Fabrica Libri Septem* (*On the Fabric of the Human Body in Seven Books*) was published in 1543, showing skeletons (with or without their muscles or circulatory systems) strolling with walking canes in the Italian countryside, casually leaning against tree trunks or columns, or even glancing down at books resting on lecterns. It was a big hit, especially with students.

After having relieved so many cadavers of their skin, Vesalius wanted to keep his own. The structure that was purported to purify the vital spirits and change them to animal spirits was simply not there. More disturbingly, the ventricles alleged to house the soul were not full of air, nor did they resemble the Church's descriptions of them. Vesalius did not question his faith or his immortal soul, but he knew that the Church fathers would if he challenged their doctrine—risky business in the era of the Inquisition. Vesalius thought that perhaps the brain and not the ventricles was where the business of the soul (sensation, understanding, and memory) was taking place. Either way, he used his head and kept quiet.

Scientists were turning the heat up a notch at the end of the sixteenth century by providing more observations. Back in Padua, Galileo was not just questioning Aristotle's (and the Bible's) notion of an Earth-centered universe but also using mathematics, measurements, and experiments to prove Aristotle wrong. The upshot was that Galileo declared that the laws of nature—that is, the laws that govern the physical world—were mathematical, which is to say mechanistic. Accused of trying to reinterpret the Bible, he was tried by the Roman Inquisition, told to shut up about the Sun, and put under house arrest.

In Paris, however, ideas were emerging. Marin Mersenne, a fellow mathematician as well as a theologian, philosopher, music theorist, and monk, supported Galileo. He lived in the convent of L'Annonciade and hosted frequent discussions in his cell with notable thinkers and scientists from across Europe. He also maintained an extensive and far-reaching correspondence with others. Mersenne had decided that if the Church was to survive the onslaught of new science and the complaints of heretics, it had to accept and absorb the view that the universe was mechanistic. God could just as easily rule over a universe that followed natural laws he had created as over a human-centered one. In fact, come to think of it, why wouldn't he in his omniscience have created a universe that could work automatically without maintenance?

Attending the sessions was another French philosopher-mathematician-scientist-priest, Pierre Gassendi. Gassendi subscribed to the notion that the world is composed of atoms, a theory first proposed in Western culture in the fifth century B.C. by Leucippus and Democritus. Atoms were described as indestructible, immutable, and surrounded by a void. Different kinds of atoms had their own specific size and shape, and all were in constant motion. Atoms could join together, and Gassendi called the resulting structure a molecule, which has a different shape and different intrinsic properties. All the macroscopic substances in the world are made up of various atoms. Gassendi did not find this belief heretical in the least. God created atoms as he did everything else.

Nonetheless, Gassendi missed the mark when he posited two types of souls. One was made up of atoms, hooked up with the nervous system and the brain, and able to perceive, feel pleasure and pain, and make decisions. There was one thing, however, that Gassendi was certain about: no atoms in any combination could reflect on themselves or perceive anything beyond what was supplied by sensation. Thus, he concluded humans must have another soul, a rational soul that was immaterial. This soul, however, was not on its own. He believed it was fused to the body during life and was dependent on the body for information from the outside world. Nonetheless, upon death, the soul proved to be immortal and fled away.

Enter the young René Descartes, philosopher, mathematician, and rationalist, who also frequented Mersenne's sessions and was all for the idea that the physical world was made up of particles and ran like a machine. A flamboyant dresser favoring taffeta, feathers, and swords, he liked to strut his stuff around Paris, which, at the time, had its own visionary version of Euro-Disney's "It's a Small World" ride in the French Royal Gardens. It was made up of water-powered automatons that moved, made sounds, and played musical instruments. The automatons were cleverly activated by the pressure exerted on the garden's tile footpaths when people walked on them. Automatons, known better today as robots, were actually rather commonplace at the time. No doubt most visitors to the garden were charmed by them.

Descartes, however, was a philosopher, one for whom a walk in the park was never just a walk in the park (hence the taffeta and feathers). He knew that these human-like robots were machines run by inanimate external forces. Yet they appeared to make rational voluntary movements. He got to thinking that certain aspects of our bodies were much the same. Our reflexes are just this: an external stimulus from the environment causes something to happen in the nervous system that results in a pre-programmed motor response. No director of action need be in charge. No soul is necessary. He also considered that a reflex response need not just be a motor response; it could instead be an emotional or cognitive one, such as a memory. Once one started down this particular garden path of thought, the theoretical possibilities of behavior generated by some sort of reflex reaction to an external stimulus were limitless. But they also were deterministic: stimulus *x* will always produce reaction *y*. Descartes gave this idea his approval for machines and animals, but did it apply to humans, too? No free will? No voluntary choices? No personal accountability for our actions? No morals or sins? Machines ourselves? That was too much.

Backing away from the void of such existential despair, Descartes began developing his history-changing idea. But the damage was already done to the study of biology. The distinguished theoretical biologist Robert Rosen points out that while no one can say what a living organism is, it is easy to say what it is like. Rosen says that Descartes got it backward: "[H]e proceeded to turn the relation between these automata, and the organisms they were simulating, upside down. What he had observed was simply that automata, under appropriate conditions, can sometimes appear lifelike. What he concluded was, rather, that *life itself was automaton-like*.* Thus was born the machine metaphor, perhaps the major conceptual force in biology, even today."[3] And born, too, the completely deterministic world that it implies.

Sure, your body will involuntarily jerk your lower leg up when you are tapped on the knee, but you can voluntarily jerk it up, too. These are two very different events, one in which your body reacts to an external stimulus and the other, according to Descartes, instituted by your mind. While the first can be described mechanically using the laws of physics in a chain of events that may lead all the way back to creation, the second, in his view, was a two-link causal chain: you will it, and presto—it happens. Why did you will it? Because you wanted to: nothing physical there to study. Just a desire. What Aristotle would dub the final cause.

Descartes rejected the idea that voluntary events were a reflex or physical mechanism that could be described scientifically. He finally came to the conclusion that while the body was governed by physical laws, human ac-

---

*Rosen's italics.

tion is caused or driven by an autonomous agent in charge, the rational soul, not made up of matter—that is, nonphysical, non-mechanistic, and not constrained by any natural laws; something from nothing. This soul was capable of consciousness, free will, abstract thinking, doubting, and morality. This is known as mind/body dualism: the idea that the body consists of physical machinery and the mind consists of nonphysical (immaterial) cognitive machinery.

Descartes was a card-carrying mathematician and scientist, and he wanted to rationally figure out the true nature of being. Since his rational mathematical approaches were working well for the physical world (he had developed analytical geometry and discovered the law of refraction, among other things), he tried to approach man's true nature using the same rational method. First, he had to chip away at everything he could possibly doubt in order to find a certainty, a foundation on which to build his arguments. It turned out that he could figure out a way to doubt just about everything, even that his mother was his mother, that the sun would rise the next day, or that he had slept in his bed in Paris the previous night rather than cavorting around Rome. He could even doubt that he had a body. After all, one's belief that one has a body is based on sensory perceptions, which are sometimes wrong. If they are wrong once, well, they could be wrong all the time. There was one thing he knew for sure, however, that he couldn't doubt. He knew for sure that he existed. In the very process of doubting, he was affirming that he was a thinking thing. Hence, *Cogito ergo sum*—I think, therefore I am.

Now that Descartes thought he had a solid foundation on which to build, he wanted to derive once and for all the true nature of being, and do it step-by-step, scientifically. He went on thinking that, because he could doubt that he had a body, he could doubt that he existed physically. From this little trail of thought he concluded, "I knew that I was a substance the whole essence or nature of which is to think, and that for its existence there is no need of any place, nor does it depend on any material thing; so that this 'me,' that is to say, the soul by which I am what I am, is entirely distinct from body, and is even more easy to know than is the latter; and even if body were not, the soul would not cease to be what it is."[4] His thinking continued in a tortuous manner, drawing conclusions from arguments that, from our current perspective, are easy to poke holes in. For example, one can easily see that just because one can doubt that one exists as a physical thing, it does not necessarily follow that one is correct and one is not a physical thing, nor that the body is not essential for thoughts. This was the shaky basis of Descartes's first argument for mind/body dualism.

Yet Descartes's arguments were without the advantages of modern knowledge. His conclusions and ideas shaped intellectual thought until modern times, and his mind/body dualism, his separation of the mind from the body and brain, has had a stranglehold on philosophers for the past 350 years. Yet at the time, his contemporaries had trouble with his conclusions. Many of his supporters, including Princess Elisabeth of Bohemia (whose correspondence with Descartes was extensive), wondered how this immaterial mind interacts with the material body. Descartes admitted to Elisabeth that he didn't have a good answer.[5] He might have been comforted to know that the question is still being batted around today. He did try, though. Descartes searched the brain and found what he thought to be the location of mind/brain interaction: the pineal gland. He wrote to her, "My view is that this gland is the principal seat of the soul, and the place in which all our thoughts are formed. The reason I believe this is that I cannot find any part of the brain, except this, which is not double."[6] You have to wonder if he wasn't grasping at straws here. After all, his search consisted of looking at calves' brains, which he already said had no immaterial soul, and at Galen's incorrect drawings.

While working all this out, Descartes slipped in the word "conscious" just once, in the Third Meditation, paragraph 32, thereby introducing the word to philosophy. As did all educated men at the time, he wrote in Latin, so he actually used the Latin word, *conscius*. Translations into French and English have not been so specific with their interpretations or the use of the word, but they employ it when Descartes himself used the verbs "to think" or "to know." Objections to its use were immediately raised. He may have been sorry he ever used it, because he continued to waffle on what he meant, vacillating as to whether consciousness is reflective—that is, a thought about a thought—or is simply thinking in general. At any rate, Descartes used the term to signify the knowledge we have of what is passing in our minds, which he argued was both indubitable and infallible, a conclusion he came to by logical reasoning. For example, if I am thinking that I have the best vineyard in the world, then I have no doubt that this is what I am thinking: it is indubitable. Also, I am not wrong that this is what I am thinking: it is infallible. Because he knew for sure what he was thinking, that meant he knew his mind better than he knew his body. For Descartes, his consciousness could not deceive him.

Descartes and the French gave birth to a philosophical industry that has striven ever since to make sense of an idea of consciousness that was never clearly defined from the start. In the end, it was not unlike the famous remark Supreme Court justice Potter Stewart made about pornography: "I shall not . . . attempt further to define [it] . . . and perhaps I could never succeed in intelligibly doing so. But I know it when I see it."

We leave seventeenth-century France equipped with a mechanistic universe and two differing descriptions of the mind. Prior to Descartes, the notion of a soul, whether material or immaterial, dominated human thought. It was as if the conscious presence we humans feel and experience makes it almost impossible to think our "soul" is a piece of flesh. Understandably, it is hard, and even downright annoying, to think that after one toils for a lifetime, the party is over at death. Aristotle tried to put us on the right course on these matters, making it clear that with death the soul died. However, even after two thousand years of human knowledge accumulation, most humans do not subscribe to the simple reality that it is our bodies (and brains) that generate what we are, in all our biological and cultural complexity.

On the road to the present, Descartes boldly separated the immortal soul (and, with it, the mind) from the mechanistic universe and mechanistic body. With the mind and flesh considered separate, the mind became the central puzzle; it was deemed immaterial, indubitable, infallible, and immutable. By promoting the mind to supernatural status, Descartes took it off the table as an object of scientific study. Descartes could never explain how this immaterial mind interacted with the material body, but his theory profoundly gummed up the thinking about the physical reality of the mind for more than two hundred years. Many of his brilliant contemporaries, such as Pierre Gassendi, agreed that there was an immaterial rational soul because they were certain that no atoms in any combination could reflect on themselves or perceive anything beyond what was supplied by sensation. As strange and useless as these seventeenth-century ideas were at the time, the idea that mental states do exist is alive and well in twenty-first-century science. Instead of an immaterial mind floating around with each of us, modern science has moved the mind into the brain and made it very physical. The question that remains is: How on earth does that work?

2.

THE DAWN OF EMPIRICAL THINKING IN PHILOSOPHY

"I don't think—"

"Then you shouldn't talk," said the Hatter.

—Lewis Carroll, *Alice's Adventures in Wonderland*

A CROSS THE ENGLISH CHANNEL from Descartes and his fellow Parisians, the British were also puzzling over the meaning of life, soul, and mind. The word "consciousness" caught on with British philosophers, and fifty years after Descartes coined it in his *Meditations*, John Locke expanded on it, as did the Scotsman David Hume. The philosophers were not alone in this endeavor. The medical world, with its interest in the body and anatomy, also began to explore the issues of mind and brain. Thomas Willis and William Petty were hard at work in Oxford, and their findings were about to influence the simmering debate on mind/brain. To some extent it was the same old story. The scientists were children of religion. Their new scientific knowledge of the world conflicted with their heartfelt childhood religious beliefs. They were experiencing what is now known as cognitive dissonance, the mental discomfort one feels when simultaneously holding two or more contradictory beliefs, ideas, or values. As a result, to reduce this discomfort, people try to explain or justify the conflict, or instead they actually change their beliefs. At the time in question, almost everyone had an overwhelming desire that the belief in God not be a casualty of the discoveries of their young science. Thus, in order to reconcile their thoughts concerning the mind, about which they knew little, and their thoughts about the body, about which they were learning more and more, these scientists began to make rather preposterous suggestions regarding how the two were intertwined. In fact, in the beginning, the neuroscientists of the era were as befuddled as the philosophers by their own felt sense of consciousness and their new commitment to objective thought.

Adding to the fount of ideas springing up inside France and England was the avalanche of work by the Germans. From Leibniz to Kant, the continent was abuzz with ideas about the nature of mind. Watching the ideas form, morph, and change is a wonder in itself. Descartes, with all of his brilliance and confidence, had thrown down the gauntlet by proposing that the mind is not made of the same stuff as the brain. This intellectual act proved challenging to the sophisticated and probing minds of the next two hundred years. In many ways the long discussion was a free-for-all, and dazzling in its importance.

The Blank Slate, Human Experience, and the Beginnings of Neuroscience

The mid-seventeenth century found England embroiled in a vicious civil war over religion and the power of the monarch. Thomas Hobbes, a royalist, and a polymath if there ever was one, had returned to Paris from London, where he had been beleaguered by the reception of a short book he had written about the politics of the time. In Paris, he found a job as tutor to the exiled Prince Charles (the future Charles II) and quickly enough became one of the guests in Mersenne's salon. From the beginning Hobbes, who was trained in physics, didn't seem concerned about the immaterial soul. He flat-out rejected Descartes's notion of the soul, which he thought was a delusion. Reason, Hobbes thought, is not enabled by some mysterious non-substance. It is merely the body's ability to keep order in the brain. Hobbes thought like an engineer: build it, make it work, and that's it—no ghosts in the system.

Hobbes had his hands full tutoring the prince and writing two books: one on vision and another about the body and its machinery. He needed an assistant and took on the clever young English medical student William Petty. For some reason, Hobbes had the preconceived notion that the senses produce a pressure that causes the motion of the beating heart. The young Petty helped him study Vesalius's books, and Hobbes found no evidence there for his theory. Nonetheless, and in keeping with the basic nature of many scientists, Hobbes plodded on.

Hobbes attended dissections with Petty, expecting to see nerves sprouting from the heart like spines from a sea urchin and spreading in all directions. They weren't there. When the penny finally dropped, Hobbes actually abandoned his hypothesis. In science, as in life, our social environment provides the opportunity for ideas to be shared back and forth. Hobbes's reversal of thought so impressed Petty that he took up this method of extensively exploring a question and being flexible: if his suppositions didn't match up with observations, he would change his mind. When he returned to England, Petty carried a material gift from Hobbes under his arm—a microscope—and a conceptual bequest in his head: the conviction that the body was an assembly of parts that ran like a machine. Still, Hobbes's greatest gift to Petty was to urge him to understand the value of answering a question through observation and experimentation, rather than twisting observations to fit one's own theories. This is easier said than done, believe me! No one likes to admit they have been wrong.

Petty became an exceptional anatomist. Not long after he returned to England, he was ensconced at Oxford. Like Vesalius before him, he had a steady stream of cadavers from the gallows. He was joined by another young physician, Thomas Willis, a royalist and a staunch Anglican. Because of this stance, which was not locally popular, Willis's training had been haphazard. Petty corrected that with a vengeance, and over the next five years he turned Willis into another extraordinary anatomist who likewise favored learning from observation and experimentation. The very young field of neuroscience was just getting started in Britain when these two took the reins. Soon, no one could ignore the centrality of the brain when it came to thinking about mental states, consciousness, and still, for some, the soul.

One Small Step for Science

Lots of things go into establishing a great scientific reputation, especially when a field is young and untested. Good luck befell Petty and Willis when, about a year after Petty started his job, a coffin arrived in his office holding the latest victim of the gallows, Anne Greene. She had been raped, and later condemned to death for murdering her newborn infant. She had hung by her neck for a full half hour, and as was common at the time, her friends had clung to her body as she swung from the rope, using their weight to hasten her death. At the autopsy the next day, Petty's office soon filled with a crowd.

With Petty at the helm, dissecting had become a spectator sport of sorts. But before he entered the room someone raised the lid of the coffin, and a gurgle was heard from inside, à la Edgar Allan Poe. A spectator was stomping on Greene's chest as Petty and Willis walked in the door. They worked frantically to revive her by various means, and succeeded to the point that she was asking for beer the next morning. The court justices wanted her to hang again, but the two doctors convinced them that she had had a miscarriage (she had been only four months pregnant) and the baby had actually been dead when it was born. She was acquitted, and later went on to have three children. This rather spectacular event brought fame and fortune to Petty and Willis. It set them up for an enviable research life with no need to go begging for financial backing.

Petty and Willis worked together for another four years. Under Petty's tutelage, Willis began to autopsy his patients when they died, to better understand the body and how it was affected by different diseases, and perhaps to find out what caused them. Later, Petty left for greener pastures, traveling as a physician with Oliver Cromwell's army in Ireland (and still later becoming a well-known economist, a member of Parliament, and a charter member of the Royal Society). Willis took over and became particularly interested in the brain, developing dissection techniques that allowed him to see its anatomy more clearly than had his predecessors. With Christopher Wren, who, among other achievements (he was an astronomer, a surgeon, and an architect), pioneered the art of injecting dyes into veins, Willis outlined the vascular system of the brain by injecting ink and saffron into the carotid artery of a dog. He was the first to understand the function of the vascular structure at the base of the brain, which, in his honor, is called the Circle of Willis.

Together, Willis and Wren produced the most accurate drawings of the human brain to date and published them in a book, *The Anatomy of the Brain and Nerves*. It sold out, and went through four printings the first year. Its anatomical drawings remained unsurpassed for more than two hundred years.

Even with all his anatomical knowledge, Willis clung to the idea that there were vital and sensitive spirits keeping the body alive, an idea from the past that seemingly wouldn't die. But Petty had trained him well. Willis eventually changed his mind when his students convinced him, through numerous clever experiments, that spirits were not involved. The blood was picking up something from the air and delivering it to the muscles, and that was the body's driving force. They didn't quite come up with the chemical element oxygen, but they were nearly there.

After carrying out numerous animal dissections, Willis saw close similarities between human brains and animal ones. From his observations, he concluded that human souls and animal souls were much the same, and differed in ways that he could observe only in their bodies. For example, animals with a bigger olfactory bulb were better at smelling. Willis saw that humans had a much bigger cerebral cortex than other animals and concluded that it was the location of memory, because humans could remember so much more. While this might seem like crude and simplistic thinking, Willis's conclusion is not a whole lot different from some of the most promising ideas floating around modern neuroscience. Indeed, the 2016 Kavli Prize in neuroscience was awarded to Michael Merzenich, the scientist who demonstrated how brain areas associated with particular activities enlarged with use.

Still, Willis's animal dissections had presented him with a big problem. Although humans are able to think in vastly different ways than other animals, their brains appeared to be very similar in organization. Since he could find no material brain substance that could account for this difference, his logical conclusion was that something else must be giving humans this ability: a rational soul. So here we go again. Since he couldn't physically locate rational thought in the body, he agreed with Gassendi's view that it was immaterial, yet located in the brain, just like Descartes claimed. Willis thought that nerves pick up sensations from the outside world and animal spirits carry them back to the brain. The spirits follow pathways deep into the brain to a central meeting place: the huge nerve bundle that connects the two half brains, the corpus callosum. Thus, once again a brilliant mind got the key issue wrong. It is as if a modern scientist looked inside a computer, didn't see anything special, and concluded there must be an immaterial spirit hovering over the motherboard that makes the computer work.

Soul as King, Not CEO

Willis, the royalist, saw this rational soul as the "king" of the body. Like the head of any big organization, the king only has information that is brought to him and does not have direct knowledge of the outside world. As with any such arrangement, this information could be flawed or could become unavailable. Because the brain itself is a physical organ, it or its parts could become ill and not provide proper intelligence, thus affecting the supply line of information to the rational soul. When the brain is ill, there is a chance that the rational soul could be affected, sometimes permanently. This was and is a powerful insight. As you might expect, Willis described various mental illnesses that he had come across in his patients to back up his theories.

Willis is important in our consciousness quest because he was one of the first to link specific brain damage to specific behavioral deficits, and because he recognized that specific parts of the brain accomplish different tasks. In his book that presented these ideas, *Two Discourses Concerning the Soul of Brutes*, he described a brain autonomously performing various tasks, not in a single location, but distributed across its terrain; he described communication channels, though in these days before electricity had been discovered, he envisioned spirits flowing within them rather than an electrical signal. He set the wheels rolling toward what has become today the field of cognitive neuroscience, the modern science that has taken up the charge of trying to understand human conscious experience.

The Blossoming of John Locke

Both Willis's empirical work on anatomy and his theorizing are thought to have had a big impact on the budding philosopher John Locke. He, too, started out as a physician, training at Oxford with Willis. One of the few series of lectures Locke felt were worth attending was Willis's lectures on anatomy. Locke subsequently became friends with another physician, Thomas Sydenham, who had also been a schoolmate of Willis's. Sydenham had gained much of his medical knowledge through real experience, the sort of method Locke came to believe was fundamental.

As he churned through hundreds of patients, Sydenham noticed that particular diseases had the same cluster of symptoms no matter who the patient was, whether a blacksmith from Sussex or the Duke of York himself. He came to believe that diseases could be differentiated from each other by their characteristic lists of symptoms, and began to classify diseases as if they were plants. This was revolutionary because up until that time, Galen's diagnoses and treatments, which were more subjective, were still popular. For Galen, each person's disease was caused by a unique imbalance of humors requiring tailor-made treatments. Sydenham, on the other hand, began what now sounds like the early rumblings of evidence-based medicine. He tried different treatments on a group of patients for a disease and evaluated and adjusted the medicines according to their effectiveness. Ironically, Galen's view is more consistent with contemporary medicine's new enthusiasm for personalized treatments, while Sydenham's view is consistent with algorithmic medicine, which is dictated by adopting standard practices and procedures for all patients with the same particular syndromes. We are going to find out in chapters 7 and 8 that it is common in science to come across differing approaches, such as these, that result in major battles over "either/or" responses when, in fact, there is at least one additional option. These are known as "false dilemmas," a type of informal fallacy. Perhaps unexpectedly, we are going to find out that it is physicists who have put their foot down and shown that neither answer alone is usually sufficient. This insight will come in handy as we consider how neurons make minds.

As you would expect from a future philosopher, Locke grilled Sydenham mercilessly about his methods. Did you really need to know the primary cause of a disease to treat it? Was it even possible to know it? While Willis thought causes were knowable and pursued this aim through his dissections and experiments, Locke and Sydenham did not. They came to believe that the causes of disease were beyond human understanding, and Locke later became convinced that the workings of the mind and the essences of things were equally unfathomable. He would approach philosophy in the same manner that Sydenham approached disease: he limited himself to talking about ideas based on everyday experiences. It is no wonder, given this stance, that Locke came up with the idea of the blank slate, the famous *tabula rasa*, on which the mind forms only from experience and self-reflection. This formulation serves as the basis for social science's current standard theory of man: nurture is in charge.

While both Locke and Descartes wound up being dualists, they differed on many details. Approaching the question of the soul from a psychological perspective, Locke wrote, "Consciousness is the perception of what passes in a Man's own mind."[1] This perception or awareness of a perception is accomplished, according to Locke, by an "internal sense" that he called "REFLECTION, the ideas it affords being such only as the mind gets by reflecting on its own operations within itself." Locke even goes so far as to say that the existence of unconscious mental states is impossible: "It [is] impossible for any one to perceive, without perceiving, that he does perceive."

Contrary to Descartes, Locke severed the connection between the soul and the mind (the thing that thinks). Remember that for Descartes, the mind and the soul are one: Thought is the main attribute of minds, and a substance cannot be without its main attribute. Thus, minds think all the time, even while sleeping, though those thoughts are immediately forgotten. Locke agreed that the waking mind is never without thoughts, but, drawing from his experience, he rejected the notion that the dreamless sleeper has thoughts. The sleeper, however, must still have his soul; otherwise, what would happen if he died while sleeping? So for Locke, the mind (with its property of consciousness) and the soul must be separate! Simple!

Descartes also limited the contents of consciousness to the present operations of the mind. Locke put on no such limits. For him, one can be conscious of past mental operations and actions. Locke saw consciousness as the glue that binds one's story together into one's sense of self, one's personal identity. He believed that consciousness allows us to recognize our past experiences as belonging to us. While he agreed with Descartes that humans have free will, he skirted the issue about how matter could produce it by adding an omnipotent God to the equation and saying God made it so.

Here were some of the smartest men in the world modeling how body, mind, and soul must work. They had to fit what seemed to them to be indisputable realities into a model that differentiates humans from animals, gods, minds, and consciousness. This is what modelers do, even today. They set up a model and they keep tweaking it, with those tweaks supposedly based on new information, until the model appears to explain a problem. In this case, the model was a mess.

At the end of the seventeenth century, ideas about what consciousness is were plentiful but confusing. Facts were accumulating that someday would provide future theorists with frameworks, but a basic, comprehensive conceptual structure explaining consciousness was still missing. In short, philosophers were at odds over the whole idea, and some thought the philosophical conception of consciousness was incoherent. It took a precocious Scottish philosopher, David Hume, who, at the age of eighteen, was already impatient with philosophy's "endless disputes," to set the idea of consciousness on a straight course toward the future. He found the ancients' moral and natural philosophy "entirely hypothetical, & depending more upon Invention than Experience."[2]

Getting Ready for Modern Conceptions

Hume seemed to burst onto the scene as a prefabricated iconoclast, ready to cut through all the talk about idealism. He thought the idea that the mind was somehow supernatural to the body was a delusion, and downright silly. He wanted to put that to rest and to structure a science of the real nature of life. Hume did just that, thereby redirecting human thinking about the nature of mind for centuries to come.

Hume soon realized that the same fallacies that were rampant in the ancient world—such as relying on hypotheses based on speculation and invention, rather than experience and observation—were also to be found among his contemporaries. Hume believed that our knowledge of reality is based on our experience and, for better or worse, on the axioms we choose. Axioms are statements that seem so evident or well-established that they are accepted without controversy or question and are simply asserted without proof. To put it simply, an axiom is a fundamentally unprovable assumption or an opinion. The problem with basing knowledge on an axiom is that then what one concludes about reality is dependent on the axioms one chooses.

As the Duke physicist Robert Brown warns, "There is nothing more dangerous or powerful in the philosophical process than selecting one's axioms. . . .
There is nothing more useless than engaging in philosophical, religious, or
social debate with another person whose axioms differ significantly from
one's own."[3]

Indeed, Hume concluded that many of the questions that philosophers
asked were pseudo questions—that is, questions that cannot be answered
with the likes of logic, mathematics, and pure reason because their answers
will always be founded at some level on an unprovable belief, on an axiom.
He thought that philosophers should stop wasting everyone's time writing
copiously about pseudo questions, dump their a priori assumptions, and
rein in their speculations, as the scientists had. They had to reject everything that was not founded on fact or observation, and that included elimi-
nating any appeal to the supernatural. Hume was pointing at Descartes
and others who believed they had conclusively demonstrated the dualist
philosophy through reason, math, and logic. Today, Hume's stance is relatively common, in part because contemporary academic philosophers are
employed by modern research universities and are surrounded by scientific experiments. Even though Cartesian ideas are still around, they are not taken
seriously by most philosophers or scientists. But in the early eighteenth
century, Hume's attack on Descartes was bold and groundbreaking.

Hume's grand plan was to come up with a "science of man": that is, to
figure out the fundamental laws guiding the mind's machinations consistent
with what was known about the Newtonian world, using Newton's scientific
method. He felt that understanding human nature, including its capabilities and frailties, would allow us to better comprehend human activities in
general. It would also allow us to appreciate the possibilities and pitfalls of
our intellectual pursuits, including what aspects of our thinking might constrain our attempts to understand ourselves. In fact, he thought his science
of man should take top billing, above Newtonian sciences, writing, "Even
Mathematics, Natural Philosophy, and Natural Religion, are in some measure dependent on the science of Man; since they lie under the cognizance of

men, and are judged of by their powers and faculties. 'Tis impossible to tell what changes and improvements we might make in these sciences were we thoroughly acquainted with the extent and force of human understanding."[4] As they say, Hume was on it, and he was bringing to the mind/brain muddle some hard-nosed clarity. Some consider him to be the father of cognitive science.

It was 1734, and, at the age of twenty-three, Hume attended Descartes's alma mater, the Jesuit College de la Flèche in Anjou, France. In his spare time, he wrote the classic *A Treatise of Human Nature: Being an Attempt to Introduce the Experimental Method of Reasoning into Moral Subjects*, which he recast, buffed up, corrected, and clarified in the 1748 publication of *An Enquiry Concerning Human Understanding*. To get going, he divided all mental perceptions into two categories: *impressions*, which are either external sensations or internal reflections, such as desires, passions, and emotions; and *ideas*, which are from memory or imagination. He argued that ideas, in the end, are copied from impressions, and he used the word "consciousness" to mean thought. Known as the copy principle, this is Hume's first principle in the science of human nature: "That all our simple ideas in their first appearance are deriv'd from simple impressions, which are correspondent to them, and which they exactly represent."[5]

Yet, Hume notes, our ideas do not occur randomly. If they did, we wouldn't be able to think coherently. Thus he proposed the principle of association: "[T]here is a secret tie or union among particular ideas, which causes the mind to conjoin them more frequently together, and makes the one, upon its appearance, introduce the other."[6] Further, these associations follow three principles: resemblance, contiguity in time and place, and causation. Of these, Hume recognizes that causation takes us beyond our senses: it links present experiences to past experiences and infers predictions of the future. Hume concludes that "all reasonings concerning matters of fact seem to be founded on the relation of Cause and Effect."[7] This conclusion foretold the downfall of Descartes's dualism.

For Hume, causality is made up of three fundamental components: priority in time, proximity in space, and necessary connection. Hume argues that the idea of priority in time comes from observation and experience. Thus, if we say that event A caused event B, then that means that A came before B. The idea of proximity in space also comes from observation, for when we observe event B and say it is caused by event A, it is in proximity to it. When I sauté garlic in olive oil, it is my house that immediately fills with a mouthwatering fragrance—it does not happen two hours later, nor at my neighbors'. The sautéing garlic and the fragrance have to be in proximity of each other in both time and space for me to infer cause and effect.

The problem for Hume arises from the third component: the necessary connection. Hume's view was that the necessary connection between cause and effect was not something that was derived from mere thought, or what he called the relation of ideas—that is, something demonstrably certain yet discoverable independent of experience, such as $4 \times 3 = 12$. No, necessary connection required experience. For example, perhaps while I am sautéing that garlic, the phone rings. Do I conclude that sautéing garlic causes the phone to ring? No. My mind infers no necessary connection, unless it happens every single time I sauté garlic.

Consider being handed a previously unknown white powder. You would have no notion of what the effect of swallowing some might be. You might be a rocket scientist, but you would only be able to describe the powder's color, texture, odor, and—if you were a rather reckless rocket scientist, which seems an oxymoron—taste. But without actual observation or previous experience of the powder's effects, you won't know them or be able to predict them. Hume sees the idea of a necessary connection between cause and effect as an idea that forms in the mind from experience. It is not an actual feature of the external world that we derive from the senses. Hume argued that when people regard any set of events as causally connected, they are merely observing that the two events always go together: events like A are always followed by events like B, but, as they say, correlation is not causation. He reasons that through association, the impression of one event brings with it the impression of the other, and if they continue to show up together, eventually the association becomes habitual. So when we see event A, through habit we expect event B. We expect the fragrance when we sauté garlic because one always follows the other, but we don't expect the phone to ring. Here Hume is prescient of Pavlov and his conditioning experiments. Hume concludes that if this habitual linking of the two events is not something that can be observed via the senses, then the only source that we can identify as the basis of the idea of causality is this compulsive linking that goes on in our brain, producing a feeling of expectation, and that feeling is the source of the idea of causality.

So, based on repeated experience, we infer or presume by force of habit that B will, in the future, always follow A. Here we run into a logic snafu. Consider this: You may have eaten shrimp for years, and then one night, driving alone with your two-year-old to meet your father at his fishing cabin, you stop at a roadside diner for a snack. You hungrily sink your teeth into a succulent shrimp, and within seconds, your throat closes up. You gasp for breath and realize you are having an allergic reaction. Instead of postprandial bliss, you are in postprandial hell. How do we get from "I have no problem eating shrimp" to "I will not have a problem eating shrimp tonight," which usually works, but in rare cases (what essayist Nassim Taleb would call a Black Swan event) doesn't? The problem is that to make any causal inference about the future, we have to assume something: the future effects will be like the past ones. We make this assumption multiple times each day. How do we arrive at this assumption? Hume shows that we do not make it through logic or reason, because, logically, it is easy to conceive that future effects may be different. Logically, you know that you could go out to start your car and, unlike yesterday and the previous five years of yesterdays, find that turning the key causes nothing to happen. Hume also discounts our use of probable (empirical) reasoning because it is circular: it supposes that which one is trying to prove (future effects will be like the past). He concludes that our belief that nature is uniform and future effects will be like the past must be derived not rationally but psychologically, from a habit based on association.

This is really walking on the wild side, and Hume knew it. He is questioning the idea of causation and making it clear that it is an axiom, an assumption that is without the hope of proof. Hume is not actually questioning that effects have causes; he is questioning where our certitude that they do comes from. In a letter to John Stewart, a professor of natural philosophy at Edinburgh, Hume wrote in 1754: "But allow me to tell you that I never asserted so absurd a Proposition, as that any thing might arise without a cause: I only maintain'd, that our Certainty of the Falsehood of that Proposition proceeded neither from Intuition nor Demonstration; but from another Source."[8]

Hume showed that although navigating the world would be extremely tricky without it, the axiom of cause and effect cannot be proved through mathematics, logic, or reason. Hume admired Newton, but here he called into question the philosophical basis of Newton's science and, with it, the mechanistic certainty of both the world and the mind/brain. We will see that questioning mechanistic certainty is going to lead us into some interesting territory in the upcoming chapters.

Me, Myself, and Who?

Hume also had a few things to say about the self, that is, our personal identity. These he garnered from introspection, and he concluded that the self consists of a bundle of perceptions, but that there is no subject in which those perceptions appear: "For my part, when I enter most intimately into what I call *myself*, I always stumble on some particular perception or other, of heat or cold, light or shade, love or hatred, pain or pleasure. I never can catch *myself* at any time without a perception, and never can observe any thing but the perception."[9] He does admit, however, that he has to account for the fact that he does have some idea of personal identity. This he attributes to associations of perceptions. For Hume, the self is nothing but a bundle of experiences linked by the relations of causation and resemblance, made from our never-ending chain of perceptions. These our imagination bundles together to give rise to an idea of an identity. Memory then extends this idea beyond immediate perceptions, linking it to those in the past.

In other words, Hume thought that we get the idea of a permanent self-identity by our way of thinking, which confounds a succession of related objects, in this case our perceptions, with an uninterrupted and invariable object with a permanent identity, such as a chair. Hume asserted that to justify "this absurdity, we often feign some new and unintelligible principle, that connects the objects [perceptions] together, and prevents their interruption or variation. Thus we . . . run into the notion of a *soul*, and *self*, and *substance*, to disguise the variation. But we may farther observe, that where we do not give rise to such a fiction, our propension to confound identity with relation is so great, that we are apt to imagine something unknown and mysterious, connecting the parts, beside their relation."[10] Hume could be chastising from the grave the many who now think of consciousness in such terms.

But he also was shaking his finger at Descartes's *res cogitans*, or "the thing which thinks." Hume objected to the idea that the mind is a thinking thing. He saw it more as a stage with the brain providing the entertainment: "The mind is a kind of theatre, where several perceptions successively make their appearance; pass, re-pass, glide away, and mingle in an infinite variety of postures and situations."[11] Hume did not assume that a diversity of experience translates into a single unity, a subject. He reasoned that through introspection, all that we can capture about our self is a bunch of perceptions and ideas. We never catch the mind that supposedly dreams them up. For him, the self is a bundle of perceptions with no bundler, no substantial, persisting, and unchanging essence. If there is no bedrock "I," then substance dualism is false.

We leave the eighteenth century with perhaps its greatest philosopher looking askance at both Descartes and Locke. Yet Hume, too, was missing something big: the many silent minds within us that influence our behavior. Our subconscious mind, which monitors each of us, is like a hidden spy in your house. This may seem like a supernatural force, but in reality it is neural processing going on below the level of conscious awareness.

Germany and the Birth of the Unconscious Mind

It was the Germans' turn to contribute to the conversation, and they stirred up talk about unconscious mental processes. Arthur Schopenhauer, for example, wrote: "For the Zeitgeist of every age is like a sharp east wind which blows through everything. You can find traces of it in all that is done, thought and written, in music and painting, in the flourishing of this or that art: it leaves its mark on everything and everyone."[12]

Early in the nineteenth century, for the purposes of our quest, that sharp east wind was blowing out of Germany, and more specifically out of the mouth of Arthur Schopenhauer. He took the wind out of the sails of Descartes, who thought that the mind is fully accessible and that nothing is hidden from conscious reflection. Schopenhauer was a philosopher who was focused on the motivations of the individual. He concluded that they aren't too pretty: humans are motivated by their will and not by their intellect, though they may firmly deny it. In his 1818 publication, *The World as Will and Representation*, he came to the conclusion that "man can indeed do what he wants, but he cannot will what he wants." In essence, not only is the will (i.e., our subconscious motivations) in charge, but the conscious intellect does not realize it. Schopenhauer made this clear when describing the will as blind and strong and the intellect as sighted but lame: "The most striking figure for the relation of the two is that of the strong blind man carrying the sighted lame man on his shoulders."[13]

Schopenhauer's framing kicked the problem of consciousness onto a much larger playing field. The mind, with all of its rational processes, is all very well but the "will," the thing that gives us our "oomph," is the key: "The will . . . again fills the consciousness through wishes, emotions, passions, and cares."[14] Today, the subconscious rumblings of the "will" are still unplumbed; only a few inroads have been made. As I write these words, enthusiasts for the artificial intelligence (AI) agenda, the goal of programming machines to think like humans, have completely avoided and ignored this aspect of mental life. That is why Yale's David Gelernter, one of the leading computer scientists in the world, says the AI agenda will always fall short, explaining, "As it now exists, the field of AI doesn't have anything that speaks to emotions and the physical body, so they just refuse to talk about it." He asserts that the human mind includes feelings, along with data and thoughts, and each particular mind is a product of a particular person's experiences, emotions, and memories hashed and rehashed over a lifetime: "The mind is in a particular body, and consciousness is the work of the whole body." Putting it in computer lingo, he declares, "I can run an app on any device, but can I run someone else's mind on your brain? Obviously not."[15] Picture Shaquille O'Neal and Danny DeVito trading brains. Danny would be ducking under doorframes and Shaq would be missing the basket by miles.

The will, according to Schopenhauer, is the will to live, a drive that wheedles humans and all animals to reproduce. For him, the most important purpose of human life is the ultimate end product of a love affair, offspring, because it determines who makes up the next generation. Schopenhauer puts the intellect in the backseat. It isn't the driver of behavior and also isn't privy to the will's decisions; it's just an after-hours spokesperson, making up stories as it goes along to explain ex post facto what the will has wrought.

Schopenhauer, in deposing conscious intellect, also opened up the Pandora's box of the unconscious. He described our distinctly conscious ideas as merely like the surface of a pool of water, while the depths are made up of indistinct feelings, perceptions, intuitions, and experiences mingled with our personal will: "Consciousness is the mere surface of our mind, and of this, as of the globe, we do not know the interior, but only the crust."[16] He said that our real thinking seldom takes place on the surface, and thus can rarely be described as a sequence of "clearly conceived judgments."

Schopenhauer ushered in the world of unconscious mental processes several decades before Freud took over the limelight, but it was by no means a new idea. Recall that Galen had recognized that many of the body's processes are carried out without cognition—in particular, those that keep the body alive, such as breathing, and also one's natural urges. In the nineteenth century, however, the idea picked up momentum. In 1867, after many years studying the physiology of the eye, the German materialist physician, physicist, and philosopher of science Hermann von Helmholtz proposed that unconscious inference, that is, an involuntary, pre-rational, and reflex-like mechanism, is at work in visual perception: the visual system in the brain takes the incoming raw visual data and stitches it together into the most coherent picture.[17] This was a different type of processing than Hume had proposed in his copy principle, but it was not a new idea, either. It had been suggested in the eleventh century by the Arab scientist Alhazen.

Helmholtz was mentor to fellow materialist physician and physicist Ernst Brücke. Both were dedicated to the idea that the elements that made up the mind are physical, and all the causal relations between the elements are governed by the same mechanical principles that govern physics and chemistry. No vital spirits, no mysticism, no ghosts. The mind and the body are one. Brücke went on to become a professor of physiology at the University of Vienna, where he would have a great deal of influence on one of his students: Sigmund Freud. Can you imagine the intense excitement of the intellectual and scientific atmosphere? No more spooks in the system. It was just the brain, made up of parts, many of which worked outside conscious awareness, all driven by chemistry and physics.

In 1868, a Dutch ophthalmologist, Franciscus Donders, came up with an idea that was going to give those interested in studying the mind's functioning a new tool. Donders realized that by measuring differences in reaction times, one can infer differences in cognitive processing. He suggested that the amount of time it takes to identify a color is the difference between the time it takes to react to a particular color and the time needed to react to light. With this idea, psychologists recognized that they could study the mind by measuring behavior, and the field of experimental psychology was born. Indeed, this very method of Donders, as well as his groundbreaking insights into cerebral oxygen consumption, led to the dramatic breakthroughs in understanding cognitive processes using brain imaging as first carried out by Marcus Raichle, Michael Posner, and their colleagues at Washington University in Saint Louis more than a hundred years later.

Rumblings about the deep unconscious mind were also being heard in England, and were already accepted by 1867, as evidenced in the writings of the British psychiatrist Henry Maudsley: "The preconscious action of the mind, as certain metaphysical psychologists in Germany have called it, and the unconscious action of the mind, which is now established beyond all rational doubt, are assuredly facts of which the most ardent introspective psychologist must admit that self-consciousness can give us no account."[18] Maudsley goes on to state that "the most important part of mental action, the essential process on which thinking depends, is unconscious mental activity."[19]

Soon after, in 1878, the publication of the British journal *Brain* was inaugurated. The next year, the journal published an article by the polymath Francis Galton in which he wrote about the findings of an experiment he performed on himself. He looked at a word written on a card and timed with a stopwatch how long it took him to associate two ideas with the word, which he then wrote down. He had seventy-five words and he performed this task in four very different locations at intervals of about a month. His findings surprised him. From his list of seventy-five words, gone over four times, he had produced only 289 different ideas. Almost 25 percent of the time a word brought forth the very same associated words during all four sessions, and in an additional 21 percent of trials, the same associations popped up on three out of the four occasions, showing much less variety than he expected. "The roadways of our minds are worn into very deep ruts," Galton remarked. He concluded:

> Perhaps the strongest of the impressions left by these experiments
> regards the multifariousness of the work done by the mind in a state
> of half-unconsciousness, and the valid reason they afford for
> believing in the existence of still deeper strata of mental operations,

sunk wholly below the level of consciousness, which may account for such mental phenomena as cannot otherwise be explained.[20]

Concurrently, the conscious mind was about to get its own field. In 1874, a young German professor of physiology, Wilhelm Wundt, published the first textbook in the field of experimental psychology, *Principles of Physiological Psychology*. In it, he marked out the territory for this new discipline, which included the study of thoughts, perceptions, and feelings. Wundt was particularly interested in analyzing consciousness, and thought that this should be the focus of psychology. He outlined a system to investigate the immediate experiences of consciousness through self-examination. This was to include an objective observation of one's feelings, emotions, desires, and ideas. Five years later, at the University of Leipzig, he opened the first psychology laboratory, thereby earning himself the moniker "father of experimental psychology." He argued that law-like regularities in humans' inner experience could be identified through experimentation. Wundt believed that neurophysiology and psychology studied the same process from different perspectives, one from the inside and the other from the outside.

Freud, the Unconscious Mind, and His Flip-Flop on Mechanism

Meanwhile, the somewhat revolutionary idea of the unconscious mind really caught some traction with the contributions of Sigmund Freud. Was it the shock factor of Freud's psychoanalytic theories that brought them to prime time? At any rate, early in his career Freud was already biting off more than he could chew. In 1895, he published *The Project for a Scientific Psychology*, in which he championed the ultra-materialist idea that every mental event was identical to a neurological event. He declared that the first step toward the goal of a scientific psychology was to identify and precisely describe the neural event associated with each mental event—an early version of the current quest to find the neural correlates of consciousness. If that weren't enough, he went on to propose the second step, "eliminative reductionism":

the vocabulary used to describe mental states was to be axed, and a new neurological vocabulary substituted. So instead of speaking of your jealousy, you would instead comment that your area J2 is firing at a specific rate and velocity. Freud proposed this change not just for those studying the brain, but for everyone. Everyone. Poetry would be quite different, as would Valentine's Day cards: "My p392J fires 95 percent faster when my L987T corresponds with your face." Probably the number of coordinates for "my" would have taken up too much space.

But just as his book was hot off the press, Freud changed his mind totally and completely. Owen Flanagan reports, "In 1895, the very year of the *Project*, Freud stated it was a 'pointless masquerade to try to account for psychical processes physiologically.'"[21] Not only that, he decided that mental events should be spoken of only in the vocabulary of psychology. No reductionism. Flanagan traces one of the roots of this argument to the German philosopher, psychologist, and former priest Franz Brentano, one of Freud's medical school instructors.

Brentano wanted philosophy and psychology to be practiced with methods as exacting as those used in the natural sciences. He distinguished between two different approaches to psychology, what he called a genetic approach and a descriptive approach. Genetic psychology would study psychology from the traditionally empirical third-person point of view, while descriptive psychology, which he sometimes referred to as "phenomenology," was geared at describing consciousness from the subjective first-person point of view. He agreed with the eighteenth-century philosophers that all knowledge was based on experience and argued that psychology must employ introspection in order to empirically study what one experiences in inner perception. Herein lie the roots of another definition of the word "consciousness": the awareness and subjective feel of phenomenal experience, which we will return to in the next chapter.

Brentano maintained that the difference between mental phenomena and physical phenomena is that physical phenomena *are the objects* of external perception, whereas mental phenomena have content and are always "about" something, that is, they are *directed at* an object. That object, Brentano specifies, "is not to be understood here as meaning a thing,"[22] but is a semantic object. So while you could desire to see a horse, you could also desire to see a unicorn, which is an entirely imaginary object, or you could desire forgiveness, which although it may or may not be imaginary, is a semantic object but certainly not an object you can put on the table. Brentano argued that this "aboutness" is the main characteristic of consciousness and referred to the status of the objects of thought with the expression "intentional inexistence." Owen Flanagan writes, "This view, which has come to be known as 'Brentano's thesis,' implies that no language that lacks the conceptual resources to capture the meaningful content of mental states, such as the language of physics or neuroscience, can ever adequately capture the salient facts about psychological phenomena."[23]

Freud, drawing on experiences he had with patients and linking them with the notion of unconscious processes, built a systematic psychological theory. He divided the mind into three levels: the *conscious mind*, which includes all that we are aware of; the *preconscious mind*, containing ordinary memory, which can be retrieved and shuttled to the conscious mind; and the *unconscious mind*, the home of feelings, urges, memories, and thoughts that are outside conscious awareness. The notion that processes involved with emotions, desires, and motivations are inaccessible to conscious reflection was not new. This idea had been tossed about not only by Descartes but also, even earlier, in the fourth century by Augustine and in the thirteenth by Thomas Aquinas, and after Descartes by Spinoza and Leibniz. Freud differed, however, in that he considered most of the contents of the unconscious unseemly. And according to his theory, the unconscious influences nearly all of our thoughts, feelings, motivation, behavior, and experiences.

Oddly, Freud, who championed the idea of a scientific psychology, never allowed his psychoanalytic theories to be tested empirically through the newly developing field of experimental psychology. While some of Freud's propositions have withstood empirical analysis—for example, it is widely accepted today that most cognitive processes are performed unconsciously—his original theories of psychopathology have not withstood close scrutiny and have generally been consigned to the trash bin.[24]

Darwin's Challenge to All

Along with the gradual acceptance of unconscious mental processing, the nineteenth century saw another big idea burst onto the scene, this time out of the British Isles, with Charles Darwin's publication of *On the Origin of Species* in 1859. The first editions flew out of bookstores and quickly aroused international interest. In the conclusion of the book, Darwin also separated himself from the mind/body dualists when he wrote: "In the distant future I see open fields for far more important researches. Psychology will be based on a new foundation, that of the necessary acquirement of each mental power and capacity by gradation. Light will be thrown on the origin of man and his history."[25] While there was some tut-tutting going around initially, by the time he published *The Descent of Man* in 1871, Darwin's grand theory of evolution through natural selection was well accepted by the scientific community and much of the general public. In that book, after detailing numerous examples of the continuity of physical and mental attributes that animals and men share, he concluded, "The difference in mind between man and the higher animals, great as it is, certainly is one of degree and not of kind."[26] Not one to endow animals with immortal souls, Darwin was again arguing against mind/body dualism.

As has been oft noted, Darwin was a soft-spoken man, and almost apologetic for throwing a "monkey" wrench into the beliefs of many people, including those of his wife. He closed *On the Origin of Species* with an upbeat, rather hopeful note for that constituency:

> Thus, from the war of nature, from famine and death, the most
> exalted object which we are capable of conceiving, namely, the

production of the higher animals, directly follows. There is a grandeur in this view of life, with its several powers, having been originally breathed by the Creator into a few forms or into one; and that, whilst this planet has gone cycling on according to the fixed law of gravity, from so simple a beginning endless forms most beautiful and most wonderful have been, and are being evolved.[27]

Darwin thought that human mental capacities must also be explained by his theory. This part of his theory was more hotly contested. It met with resistance both from traditional dualists and from the empiricist followers of Locke and Hume, the purveyors of the human brain as a *tabula rasa*, who thought that all knowledge comes from sensory experience. These disputes stifled progress on the conundrum of consciousness for many years. Eventually a different approach to psychology, with its roots in Hume's association principles, took up the reins: behaviorism.

BY THE END OF THE nineteenth century, many philosophers were insisting that the mind had to have its physical brain, which somehow contained memories and cognition. Some physiologists also required the spinal nerves, and others insisted that the body, too, was part of the package.

Locke separated the mind from the soul, and infused the mind with rational reflection, ethical action, and free will. The mind is the basis of consciousness, volition, and personhood, but it is fallible and can generate illusions and error, and its only coin is conscious ideas. Nothing bubbles up from unconscious depths. Locke skirted the issue of how matter could produce something like free will by adding an omnipotent God to the equation and saying that he made it so. Hume eliminated those supernatural powers from the equation and tried to establish a true science of human minds. In doing so, he realized the limits of the human mind, how all thought has to be constrained by its capacities. Thus, he even questioned the philosophical basis of Newton's mechanistic science as a way of looking at the world, by undercutting the foundation of humans' grasp of physical causality.

Schopenhauer insisted that unconscious motivations and intentions drive us, not conscious thinking, which makes him more of an a posteriori apologist. Helmholtz showed that our perceptual systems are not veritable Xerox machines, but rather stitch perceptual information together in a best-guess sort of way. Then along came Darwin, who plopped our brains down on an evolving continuum, instructing us to use natural selection to figure out how they got the way they are and to leave God out of it.

So we enter the twentieth century, still confused, still with the same questions, but with a couple of new strategies to employ: study the difference in reaction times to particular tasks, and employ a new descriptive psychology focusing on the subjective first-person point of view. The next one hundred years were surely going to be full of new insights, new scientific findings, wholly new ways of thinking about consciousness. Humankind cracked the atom, cracked the code of DNA, went to the moon, and could now take pictures of the living human brain. Surely, something had to give on the problem of consciousness.

3.

TWENTIETH-CENTURY STRIDES AND OPENINGS TO MODERN THOUGHT

There are some people, nevertheless—and I am one of them—
who think that the most practical and important thing about
a man is still his view of the universe. . . . We think the question
is not whether the theory of the cosmos affects matters, but
whether in the long run, anything else affects them.

—G. K. Chesterton

A T THE START of the twentieth century, the philosophy of mind and brain was still divided into two battling camps: the rationalists and the empiricists. At the end of the century, as we shall see, things were not a whole lot better. It is almost as if our human brains have a limited set of ideas, and whatever the scientific data or intellectual mood of the day might be, one of these two views is trotted out. But back to the beginning of the century. It was time for the upstart Americans to brashly chime in, and William James was the first to give the issue of consciousness a good hard look. In 1907, he gave a series of lectures at Harvard and began with the above quote from G. K. Chesterton, which neatly summarizes the great mind/

brain question of philosophy: Can a mind state—an immaterial belief, an idea—affect matter, that is, a brain state?

James agreed with Chesterton that this was the important question. The topic of his lectures was a new philosophical method: *pragmatism*, the brainchild of James's friend Charles Peirce, which grew out of discussions they had had with other philosophers and lawyers at the Metaphysical Club, a short-lived but influential intellectual salon they cofounded in Cambridge, Massachusetts, in the 1870s. Pragmatism didn't get much attention until James further developed and promoted it twenty years later. In the first lecture, James pointed out what was hidden in plain sight, that philosophers and their philosophical stances are biased by their temperaments:

> The history of philosophy is to a great extent that of a certain clash of human temperaments. . . . Of whatever temperament a professional philosopher is, he tries when philosophizing to sink the fact of his temperament. Temperament is no conventionally recognized reason, so he urges impersonal reasons only for his conclusions. Yet his temperament really gives him a stronger bias than any of his more strictly objective premises. It loads the evidence for him one way or the other, making for a more sentimental or a more hard-hearted view of the universe, just as this fact or that principle would. He trusts his temperament. Wanting a universe that suits it, he believes in any representation of the universe that does suit it.[1]

And here is the really great, brashly American part. James divides American philosophers into two groups, according to their temperaments: "tender-foot Bostonians" and "Rocky Mountain toughs." He sees this temperamental dichotomy not only in philosophy but also in literature, art, government, and manners. And, of course, the two have low opinions of each other: "Their mutual reaction is very much like that that takes place when Bostonian tourists mingle with a population like that of Cripple Creek.

Each type believes the other to be inferior to itself; but disdain in the one case is mingled with amusement, in the other it has a dash of fear." He further sketches out the groups: the tender-minded, tender-footed Bostonians are rationalistic (devotees to abstract and eternal principles), intellectualistic, idealistic (in the sense that they believe all things proceed from the mind), optimistic, religious, free-willist, monistic (that is, rationalism starts from wholes and universals, and makes much of the unity of things), and dogmatic. Descartes, deep down, is a tender-foot!

The tough-minded Rocky Mountaineers are quite the opposite: empiricist (lovers of facts in all their crude variety), sensationalistic, materialistic (all things are material, no immaterial mind), pessimistic, irreligious, fatalistic, pluralistic (meaning empiricists start from the parts and they make the whole a collection of those parts), and skeptical (that is, open to discussion). Hume, a tough!

But James realized that most of us are not purely one or the other:

Most of us have a hankering for the good things on both sides of the line. Facts are good, of course—give us lots of facts. Principles are good—give us plenty of principles. The world is indubitably one if you look at it in one way, but as indubitably is it many, if you look at it in another. It is both one and many—let us adopt a sort of pluralistic monism. Everything of course is necessarily determined, and yet of course our wills are free: a sort of free-will determinism is the true philosophy. The evil of the parts is undeniable; but the whole can't be evil: so practical pessimism may be combined with metaphysical optimism. And so forth—your ordinary philosophic layman never being a radical, never straightening out his system, but living vaguely in one plausible compartment of it or another to suit the temptations of successive hours.[2]

Yet the more philosophically minded "are vexed by too much inconsistency and vacillation in our creed. We cannot preserve a good intellectual conscience so long as we keep mixing incompatibles from opposite sides of the line."

So James describes the common layman as wanting facts, science, and religion. But what philosophy was giving him was "an empirical philosophy that is not religious enough, and a religious philosophy that is not empirical enough."[3] Practical help, not highly abstracted absolutist philosophy, was needed to navigate a world whose denizens were interested in the science with which they were being bombarded but also found comfort in religion or romanticism. James thought that the pragmatic method would supply that help. Its foundation is based on the idea that our beliefs are our rules for action—that when we form a belief, we acquire a disposition to act in some distinctive way. In order to understand the significance of a belief, you simply need to determine what action that belief would produce. If two different beliefs produce the same action, then let it rest:

> The pragmatic method is primarily a method of settling metaphysical disputes that otherwise might be interminable. Is the world one or many?—fated or free?—material or spiritual?—here are notions either of which may or may not hold good of the world; and disputes over such notions are unending. The pragmatic method in such cases is to try to interpret each notion by tracing its respective practical consequences. What difference would it practically make to any one if this notion rather than that notion were true? If no practical difference whatever can be traced, then the alternatives mean practically the same thing, and all dispute is idle. Whenever a dispute is serious, we ought to be able to show some practical difference that must follow from one side or the other's being right.[4]

Although pragmatism is based on the idea that a mental state could be the cause for action, it is a method only and does not advocate particular results. It is open to various types of methods that are employed in different sciences. It is a method, however, that rejects a priori metaphysics and endless intellectualist accounts of thought. It was appealing to stimulus-response psychologists, followers of Hume's theories of association, who dominated the new field of experimental psychology launched by Wilhelm Wundt, and developed further and brought to New York by his student, the charismatic psychologist Edward Titchener. Another particularly influential character was Edward Thorndike. In his 1898 monograph, *Animal Intelligence: An Experimental Study of the Associative Processes in Animals*, he formulated the first general statement about the nature of associations: the law of effect. He had noticed that a response that was followed by a reward would be stamped into an organism as a habitual response and that the response would disappear if no reward was given. This stimulus-response mechanism could potentially be the mechanism that established increasingly adaptive responses.

Stimulus-response psychology, also known as behaviorism, quickly came to dominate the studies of associative processes in America. Behaviorists approached psychology from the viewpoint that its appropriate subject matter was behavior rather than mental and subjective experience; it should be studied using methods appropriate to the natural sciences, not introspection. They thought that given a particular environmental stimulus, the behavior of any animal, including humans, could be explained by a law-like tendency to react in a certain way.

Dominating the field was the dynamic figure of John B. Watson. Watson's stance was that psychology could be objective only if it was based on observable behavior, and he rejected all talk of mental processes that could not be publicly observed: looking inside the black box of the brain was verboten. Ignoring Darwin's theory of innate mental processes, Watson became committed to the idea that everybody has the same exact neural equipment; the mind is a blank slate, and any child can be trained to do anything by learning through stimulus-response and reward. This idea appealed to the American sense of equality. Soon, most of the directors of American psychology departments held these views, ignoring Darwinian theory's assertion that complexity is built into the human organism through the process of natural selection and evolution. Behaviorism reigned in the United States for the next five decades, presided over for many years by its major spokesman, the Harvard psychology professor B. F. Skinner.

Of course, even in times when major themes dominate the academic world, there are contrarian stirrings. New methods for studying "mental" processes were being developed that not only made their way steadily into experimental psychology but also became the dominant tools of exploration in modern times.[5] Still, talk about mental states and consciousness was largely out of the question in the United States until the cognitive revolution led by George A. Miller at Harvard and the notion of mentalism led by Roger W. Sperry at Caltech reared up at midcentury.

The Canadians' Resistance and the Rise of Modern Neuroscience

Thankfully, researchers in Canada did not jump on the behaviorist bandwagon. In fact, Montreal's first neurosurgeon, Wilder Penfield, was making amazing discoveries on patients who had seizures that could only be controlled by removing the portion of their cerebral cortex that was instigating them. In order to locate the region of the seizure foci, Penfield stimulated parts of the cerebral cortex with electrical probes and observed the patient's responses. During this surgery the patients were awake, under local anesthesia only, so they could comment on what they did or did not feel. Penfield traced out, in the sensory and motor cortices, maps that corresponded to the body's parts, that is, the physical representation of the human body, located within the brain.* The body that was represented, however, was not normally proportioned. Instead, it was proportional to the degree that the particular body part was innervated: more innervation, more brain area devoted to it. Penfield, along with his close associate Herbert Jasper, a physiologist, got

---

*The presence or absence of these representations is the source of the more recently described phantom limb (presence of representation but absence of limb) or body identity integrity disorder (absence of limb representation but presence of limb, which the patient views as foreign and wants amputated).

the ball rolling when it came to understanding the localization of brain function. Penfield wrote, "Consciousness continues, regardless of what area of cerebral cortex is removed. On the other hand consciousness is inevitably lost when the function of the higher brain stem (diencephalon*) is interrupted by injury, pressure, disease, or local epileptic discharge." Yet he is quick to qualify that "to suggest that such a block of brain exists where consciousness is located, would be to call back Descartes and to offer him a substitute for the pineal gland as a seat for the soul."[6]

Penfield goes on to describe that while sensory information is processed through the diencephalon, that is, through subcortical regions, information travels back and forth between the subcortex and different areas of the cortex: "Thus, the differing processes of the mind are made possible through combined functional activity in diencephalon and cerebral cortex, not within diencephalon alone."[7] He also states that the final process that is necessary for a conscious experience is to have attention focused on the mental state that is producing it. He predicts that this process is part of the diencephalon's function. We can discern in these writings that Penfield is using the word "consciousness" to mean two different things. In the first instance, he is talking about the mental state of being alert and aware, that is, not in a coma. In the second two, he is referring to Descartes's consciousness, meaning a thought or a thought about a thought, and he adds that focusing attention is a necessary component.

Penfield added to his group a psychologist, Donald Hebb, to study the effects that brain injury produced in his patients and the results of surgery for the functioning of the brain. Like most people who study patients with brain injuries, Hebb came away convinced that the workings of the brain explain behavior. While this may seem elementary to us today, as it did to Galen so many centuries ago, mind/body dualism still held many enthralled in 1949, when Hebb published his book *The Organization of Behavior:*

---

*The diencephalon contains the epithalamus, thalamus, hypothalamus, ventral thalamus, and third ventricle.

A Neuropsychological Theory. At a time when psychology was still in the tight clutches of behaviorism, Hebb took the psychological world by storm by boldly stepping into the black box of the brain, thumbing his nose at the "off-limits" constraints imposed both by the empiricist Hume and by behaviorists. He postulated that many neurons can combine into a coalition, becoming a single processing unit. The connection patterns of these units, which can change, make up the algorithms (which can also change with the changing connection patterns) that determine the brain's response to a stimulus. From this idea came the mantra "Cells that fire together wire together." According to this theory, learning has a biological basis in the "wiring" patterns of neurons. Hebb noted that the brain is active all the time, not just when stimulated; inputs from the outside can only modify that ongoing activity. Hebb's proposal made sense to those designing artificial neural networks, and it was put to use in computer programs. By opening the black box of the brain and looking inside, Hebb had also launched the first volley that started the revolution against behaviorism.

The Cognitive Revolution in America

Behaviorism's grip on American psychology began to loosen in the 1950s, especially when a bevy of young, brilliant mind scientists, such as Allen Newell, Herbert Simon, Noam Chomsky, and George Miller, together founded cognitive psychology. Miller, for example, did what a scientist is supposed to do when presented with compelling new evidence: he changed his mind. Miller was researching speech and hearing at Harvard when he wrote his first book, *Language and Communication.* William James would have been impressed by the full-disclosure preface, in which Miller made no bones about his partiality: "The bias is behavioristic." In the section on psychology, which was about the differences in how people use language, his probabilistic model of word choice was based on a behaviorist pattern of learning-by-association. The title of a textbook he wrote eleven years later, *Psychology: The Science of Mental Life,* announced a complete dismissal of his previous stance that psychology should study only behavior. What prompted Miller's change of mind was the rise of information theory: the introduction of Information Processing Language I, a computer language that implemented several early artificial intelligence programs, and computer genius John von Neumann's ideas on neural organization, which proposed that the brain may run in a manner similar to a massively parallel computer. Parallel computing means that several programs can run at the same time, in opposition to serial programming, in which only one program can run at a time.

Perhaps for Miller the final nail in the coffin of behaviorism was meeting the brilliant linguist Noam Chomsky. Chomsky was shaking the psychological world to its very roots by showing that the sequential predictability of speech follows grammatical rules, not probabilistic rules. And these grammatical rules were shocking: innate and universal—that is, everybody has them, and they are already wired into the brain at birth. Just like that, the notion of a *tabula rasa* had to be tossed out kicking and screaming, though some of those screams can still be heard.

In September 1956, Chomsky published "Three Models for the Description of Language," his preliminary version of these ideas on syntactic theories. Taking the linguistic world by storm, Chomsky transformed the study of language in one fell swoop. Miller's takeaway from that paper was that associationism, the pet of behaviorists and of radical behaviorist B. F. Skinner in particular, could not account for how language is learned. While the behaviorists had elucidated some aspects of behavior, there was something more going on in that black box that behaviorists could not explain and never would. It was about time they started trying to figure it out.

Miller began to explore the psychological implications of Chomsky's theories with the ultimate goal of understanding how the brain and mind work as an integrated whole. At the time, however, Miller was leery of one aspect of mental life. He wrote in *Psychology: The Science of Mental Life* that for the time being, the study of consciousness needed to be put on the shelf: "Consciousness is a word worn smooth by a million tongues. Depending upon the figure of speech chosen it is a state of being, a substance, a process, a place, an epiphenomenon, an emergent aspect of matter or the only true reality. Maybe we should ban the word for a decade or two until we can develop more precise terms for the several uses which 'consciousness' now obscures."[8]

The word "consciousness," which Descartes used to mean either a thought or a thought about a thought, had blossomed over the years and taken on all sorts of additional meanings. In addition to what Miller wrote, it had also become entwined with awareness, self-awareness, self-knowledge, access to information, and subjective experience. While most researchers followed Miller's advice and set the study of consciousness on the shelf, one intrepid group did not. Instead, they did an inventory of what science could say about consciousness up to that point.

Seeking Clarity at the Vatican

While Miller was locking the word "consciousness" away, the Pontificia Academia Scientiarum (Pontifical Academy of Sciences) was bringing it front and center for a study week in 1964. The Academy traces its roots back to the Accademia dei Lincei (Academy of Lynxes), founded in 1603 by an eighteen-year-old Roman prince and naturalist, Federico Cesi, whose uncle was a well-connected cardinal. Cesi founded the Academy to understand the natural sciences through observation, experiment, and inductive reasoning. To symbolize those goals, he chose the sharp-eyed lynx as the academy's emblem. In 1610, Galileo was named its president.

It was rocky times for such an endeavor, and the Academy did not survive Ceci's early death at age forty-five. It was resurrected in 1847 by Pope Pius IX as the Accademia Pontificia dei Nuovi Lincei (Pontifical Academy of the New Lynxes). Later, after the unification of Italy and its separation from the Vatican in 1870, the Academy of the New Lynxes split into two: the Royal National Lincean Academy under the Italian flag and another destined to become the Pontifical Academy of Sciences, re-founded in 1936 by Pope Pius XI and headquartered in Vatican City. The academy, though founded by the Pope and located within the walls of the Vatican Gardens, has no restrictions on its research. It is made up of scientists from many countries and disciplines and is charged with the goal of "the promotion of the progress of the mathematical, physical, and natural sciences, and the study of related epistemological questions and issues." In September 1964, the Pontifical Academy called for a study week on the topic of "Brain and Conscious Experience," to be headed up by the renowned physician and physiologist Sir John Eccles.

Eccles was an Australian. While in medical school, he was not only an avid student but a pole vaulter as well. Reading *On the Origin of Species* for his zoology course spurred him on to read philosophical writings, both classical and contemporary, on the mind/brain problem.[9] Medical school, however, did not provide the answers to his questions regarding the interaction of mind and body, and he resolved to become a neuroscientist.[10] He also resolved to win a Rhodes Scholarship to Oxford and work with the famed neurophysiologist Charles Sherrington. This he did. He set off in 1925 for England, half a world away.

Eccles went on to study the method of neural transmission at the synapse. Initially, he was convinced that transmission was electrical. During this period, he met and was encouraged to rigorously test his hypothesis by the philosopher Karl Popper, who emphasized that the strength of a hypothesis depended on the failure of a thorough investigation to falsify it, not on evidence that apparently supported it. Through dogged hypothesis testing, Eccles changed his mind and resolved that synaptic transmission was chemical. Such an about-face prompted a long-standing friend, Sir Henry Dale, to write, "A remarkable conversion indeed! One is reminded, almost inevitably, of Saul on his way to Damascus, when the sudden light shone and the scales fell from his eyes."[11] Over the next decade Eccles went on to elucidate the mechanisms involved in the firing and inhibition of motor neuron synapses in the spinal cord and then turned his sights on the thalamus, hippocampus, and cerebellum. The year before the pontifical conference, Eccles was awarded the Nobel Prize in Physiology or Medicine. A few years earlier, he had been honored with knighthood for the same research. He was a legend in his own time, and for those of us who knew him, he was crackling smart, endlessly energetic, and a great scientist. He also had been raised Catholic and was a declared dualist. The pragmatist William James would not have been surprised that his belief was a rule for action, and he spent a lifetime searching for mechanisms by which the mind controls the body.

In a 1951 essay in *Nature* titled "Hypotheses Relating to the Brain-Mind Problem," Eccles stated that "many men of science find in dualism and interaction the most acceptable initial postulates in a scientific approach to the problem of mind and brain. In such an approach the question arises: What scientific hypotheses may be formulated that bear in any way on the hitherto refractory problem of brain-mind liaison?"[12] He went on to propose such a hypothesis. Although he thought that every perceptual experience is the result of a specific pattern of neuronal activation, and that memory is caused by an increase in synaptic efficacy, for some reason he thought experience and memory are "unassimilable into the matter-energy system." He proposed instead that the activated cortex has "a sensitivity of a different kind from any physical instrument" and that "mind achieves liaison with the brain by exerting spatio-temporal fields of influence that become effective through this unique . . . function of the active cerebral cortex." Wow! That is basically voodoo with fancy language. He had replaced Descartes's pineal gland with the mysteriously sensitive activated cerebral cortex. Indeed, two hundred years after Descartes, Eccles continued his tradition of dualism even though he spent sixty hours a week working on and recording neurons and had otherwise totally adopted the determinist agenda. It is mind-boggling.

Part of Eccles's job description for heading up the study week was to pick the other attendees and to publish the discussions, which resulted in the landmark book *Brain and Conscious Experience*. The only bias that Eccles could be accused of was that the conference was a bit heavy in physiologists, but all tended to wear more than one hat. He did succeed in getting the top scientists in their fields, ranging across neurophysiology, neuroanatomy, psychology, pharmacology, pathology, biopsychology, neurosurgery, chemistry, communications, cybernetics, biophysics, and animal behavior. The academy, in its aim to study physical, mathematical, and natural sciences, had declared a single restriction: no philosophers. Eccles was not happy with this, but among the group were what one reviewer described as "amateur philosophers of no mean order." The reviewer went on to conclude that "as a single volume dealing with recent progress in our understanding of the cortex, [*Brain and Conscious Experience*] is probably unequalled."[13]

Prior to the meeting, the participants were given a brief by the Academy in which consciousness was described as "the psychophysiological concept of perceptual capacity, of awareness of perception, and the ability to act and react accordingly." As Roger Sperry, my mentor and future Nobel laureate, put it to me when he returned to Caltech, "The Pope said, 'The brain is yours but the mind is ours.'" The talks were loosely divided among these three aspects of consciousness: perception, action, and volition.

The zoologist of the group, William Thorpe, expanded on this:

The term consciousness, although having innumerable overtones of meaning, involves, I think, three basic components. First, an inward awareness of sensibility—what might be called "having internal perception." Second, an awareness of self, of one's own existence. Third, the idea of consciousness includes that of unity; that is to say, it implies in some rather vague sense the fusion of the totality of the impressions, thoughts, and feelings which make up a person's conscious being into a single whole.[14]

In discussing cerebral events as they relate to conscious experience, Eccles asked the question "How can some specific spatiotemporal pattern of neuronal activity in the cerebral cortex evoke a particular sensory experience?"[15] That question was left unanswered, and remains so.

In reading through the volume, I have to chuckle at the impact of Roger Sperry's talk about our split-brain research, which was then in its infancy. In his written summary, Sperry had stated: "Everything we have seen so far indicates that the surgery has left these people with two separate minds, that is, two separate spheres of consciousness."[16] The animated discussion afterward indicates how fascinating our findings were and what a showstopper his talk was. He was telling the Vatican and his colleagues that the mind could be divided into two with the slice of a surgeon's knife.

At the time, Sperry was in the midst of changing his own mind, in part because of split-brain research, and readjusting his basic stance on brain function. He was turning his back on materialism and reductionism as they were then defined and calling himself a "mentalist." Earlier that year, while working on a nontechnical lecture on brain evolution, he was shocked when he found himself concluding "that emergent mental powers must logically exert downward causal control over electrophysiological events in brain activity."[17] At the time, the notion that a mental state could affect a brain state was complete heresy in the world of neuroscience—and to a great extent it still is. I came to similar conclusions, and reintroduced the idea of mental processes having a downward causative effect in 2009 in my Gifford Lectures in Edinburgh, and I rediscovered how determinists of all stripes are not very receptive to the idea. The central premise of both behaviorism and materialism is that the objective physical brain process is a causally complete stimulus-response network within itself: it gets no input from conscious or mental forces, nor does it need any. In many ways, the book in your hands is a fresh attempt to wrestle with this problem.

At the Vatican conference, Sperry soft-pedaled his growing mentalist stance by merely saying at the close that "consciousness may have real operational value, that it is more than merely an overtone, a by-product, epiphenomenon, or a metaphysical parallel of the objective process."[18] At another point he paraphrased this as "a view that holds that consciousness may have some operational and causal use."[19]

Eccles, wearing his materialist hat, admitted, "I am prepared to say that as neurophysiologists we simply have no use for consciousness in our attempts to explain how the nervous system works."[20] He also admitted, "I don't believe this story, of course; but at the same time, I do not know the logical answer to it."[21] He kept his dualist position.

At the end of the week, the summation of the conference was handed over to the MIT psychologist Hans-Lukas Teuber, one of the founding fathers of neuropsychology. He was famous for his brilliant "wrap-ups" of conferences and scientific meetings, which he spiced with elaborate eyebrow wiggling.[22] He charted where the participants agreed and disagreed, and noted where there were gaps in knowledge, providing a concise summary of the field at the time as only he could do. The others agreed that they understood quite a bit about cortical processing of sensation and vision, and that if they understood an equal amount about motor action, memory retention, and awareness—which they didn't—then they would be very much further along in understanding conscious experience. Teuber lamented: "Every conceivable divergence of opinion seemed to arise when we tried to delineate the systems or mechanisms that might be necessary for consciousness. We were not even quite sure . . . as to how we might decide what consciousness is for."[23]

Teuber was an intense man who helped mentor me in the early days. I can remember a visit he made to Santa Barbara in the late sixties. My wife and I were holding a reception for him in our home in Mission Canyon when he winked at me and said he wanted a word with me alone. We stepped into the bedroom, whereupon he took a recent manuscript I had submitted to the journal *Neuropsychologia* out of his briefcase and began going through it with a red pencil. I was stunned but grateful for the attention. When we were finished, he jumped up and said, "Let's go rejoin the party!" I must have said something coherent, because he then invited me to join the newly founded International Neuropsychological Symposium, a wonderful organization that meets yearly in different cities of the world, an event I relished for twenty years.

The Vatican conference, of course, did not solve the mind/body problem. Even so, the varied opinions that were expressed launched a set of arguments and debates within biology and philosophy that continue to this day. The same list of possible solutions was on the table, with Eccles holding tight to Descartes's two-substance view that mind and body were two separate entities, though he never could find empirical evidence for it. Most leaned toward the materialist view that the mind—consciousness—was produced by matter, but how that happened was as puzzling as ever.

The Vatican conference was a turning point for Sperry. The possibility that mental states could causally affect brain states became his scientific passion, with all of its implications. The psychiatrist of the group, Hans Schaefer of the University of Heidelberg, subscribed to that theory, based on his belief that psychoanalysis worked. Evolutionary theories allowed materialist theories of consciousness to come in two flavors: emergentism and panpsychism. The former proposes that consciousness emerges from unconscious matter once that matter achieves a certain level of complexity or organization. Sperry was leaning heavily in this direction. The latter, panpsychism, tosses the whole problem out by suggesting that all matter has subjective consciousness, albeit in a wide range of types. The idea here is that there is no need for the idea of emergence and complexity to explain consciousness. Consciousness is a primordial feature of all things, from rocks to ants to us.

Returning from the conference, Sperry continued to refine his views. He came out of the mentalist closet the following year at a lecture at his alma mater, the University of Chicago. "I am going to align myself in a counterstand, along with that approximately 0.1 per cent mentalist minority, in support of a hypothetical brain model in which consciousness and mental forces generally are given their due representation as important features in the chain of control."[24] He explained his reasoning: "First, we contend that conscious or mental phenomena are dynamic, emergent, pattern (or configurational) properties of the living brain in action—a point accepted by many, including some of the more tough-minded brain researchers. Second, the argument goes a critical step further, and insists that these emergent pattern properties in the brain have causal control potency—just as they do elsewhere in the universe. And there we have the answer to the age-old enigma of consciousness." Sperry had taken conscious experience to be a nonreductive (it can't be broken down into its parts), dynamic (it changes in response to neural activity), and emergent (it is more than the sum of the processes that produce it) property of brain activity, and said that it could not exist apart from the brain. Denying any type of dualism, he emphasized, "The term ['mental forces'] fits the phenomena of subjective experience but does not imply here any disembodied supernatural forces independent of the brain mechanism. The mental forces as here conceived are inescapably tied to the cerebral structure and its functional organization."[25] There are no spooks in the system.

By the early 1970s, this notion was gaining some limited acceptance, and it contributed to the growing anti-behaviorism sentiment. Mental images, ideas, and inner feelings were back on the table. They even could have a causal role in explanations. The "cognitive revolution" was on, and it continues to the present.

Modern Philosophers Take a Stab

Meanwhile, philosophers were wrangling about theories that were incorporating the materialist view of the brain. After retiring in 1975, Eccles left the lab behind and joined forces with the eminent philosopher Karl Popper. They agreed with Descartes that the brain must be open to nonphysical influences if mental activity is to be effective,[26] that is, if a thought can affect a brain state. Eccles tried to come up with testable hypotheses, but he did not succeed and finally settled for a model of mind/brain interaction without any experimental evidence and without a testable hypothesis. While his form of dualism didn't have much of a fan club, a different type of dualism flared up again, fanned by the wings of bats.

New York University's well-known philosopher Thomas Nagel published a head-turning article in 1974 titled "What Is It Like to Be a Bat?" and with it ushered in the whole "But how can we explain the experience of redness?" line of argument. Arguing that consciousness has an essential subjective character (just as Franz Brentano had argued), Nagel states that "an organism has conscious mental states if and only if there is something that it is like to be that organism—something it is like for the organism." "Like" does not mean "resemble," such as in the question "What is ice skating like? Is it like roller skating?" Instead, it concerns the subjective qualitative feel of the experience, that is, what it feels like for the subject: "What is ice skating like for you?" (For instance, is it exhilarating?) Nagel called this the "subjective character of experience." It has also been called "phenomenal consciousness," and, although he doesn't say it, it is also referred to as *qualia*.

For Nagel, there is something that it *feels* like for the subject of an experience to have that experience, and there is something that it feels like for a creature to be the species it is and no other; and the subjective character of a mental state can be apprehended only by that particular subject. This idea was like (in the sense of "resemble") serving a plate of carbonara to a hungry footballer: it was gobbled up by philosophers, who had been pining, according to the philosopher Peter Hacker, for salvation from "reductive physicalism or soulless functionalism."[27] For some, the escape hatch became: Science is objective, consciousness is subjective; never the twain shall meet, or if they do, they will meet by some new, as yet undescribed physics or fundamental laws (Nagel's current stance).[28]

The philosopher Daniel Dennett, however, has become notorious for challenging Nagel's question. He says that Nagel doesn't want to know what it would be like for him to be a bat. He wants to know objectively what it is subjectively like: "It wouldn't be enough for him to have had the experience of donning a 'batter's helmet'—a helmet with electrodes that would stimulate his brain into bat-like experiences—and to have thereby experienced 'battitude.' This would, after all, merely be what it would be like for Nagel to be a bat. What, then, would satisfy him? He's not sure that anything would, and that's what worries him. He fears that this notion of 'having experience' is beyond the realm of the objective."[29]

Beyond the realm of science. This is what many consider the unbridgeable gap between subjective and objective realms. The new dualism.

Dennett handles this problem by denying it. He laments that one of the problems with explaining consciousness is that we all think we are consciousness experts, and have very strong beliefs about it, just because we have experienced it. He complains that this doesn't happen to vision researchers. Even though most of us can see, we don't think we are vision experts. Dennett claims that consciousness is the result of a bag of tricks: our subjective experience is an illusion, a very believable one, one that we fall for every time, even when it has been explained to us how it comes about physically, just like some optical illusions that still fool us even though we know how they work.

The philosopher Owen Flanagan also disagrees that there is an unbridgeable gap, writing, "It is easy to explain why certain brain events are uniquely experienced by you subjectively: Only you are properly hooked up to your own nervous system to have your own experiences."[30] This seems reasonable. So what's the big deal? While most philosophers today can accept that every mental event and experience is some physical event, many nonetheless resist the conclusion that the essence of a mental event or experience is completely captured by a description at the neural level. Flanagan simply takes the stance that there is nothing mysterious about the fact that conscious mental states possess a phenomenal side. It is all part of the coding.

So, as we glide into the modern era, nothing is resolved. While neuroscience had figured out how reflexes work, how neurons communicate with each other, how traits and much more are inherited, the field remained clueless as to how the brain creates what we have come to call our phenomenal conscious experience. Nothing like an Einsteinian moment had occurred for the mind/brain sciences, and while the contents of the black box could be explored by cognitive psychologists, young scientists were advised to leave the topic of consciousness alone.

Francis Crick to Modern Science: It's Okay to Study Consciousness

Two decades after George Miller had set consciousness aside for a decade or two, none other than the ever-intrepid, supremely intelligent, creative, and curiosity-driven Francis Crick stepped in and pulled it off the shelf. Yes, that Francis Crick. From an early age, Crick had been interested in two unknowns: the origin of life and the puzzle of consciousness. After spending thirty years on the first unknown, he was itching to tackle the second. And so in 1976, at the tender age of sixty, when most people are looking forward to retirement, he packed his bags, left Cambridge, and headed off to the Salk Institute in San Diego to begin a second scientific career in neuroscience.

I happened to be visiting the Salk soon after his arrival and was ushered into his spectacular office overlooking the sea. He was just beginning to immerse himself in neuroscience and was surrounded by other talented researchers. I hadn't a clue how to add to the conversation, so I asked him, "How does one think about the timescales that are prescribed by molecular processes, and how do they relate to the different timescales operative in neural activity? Each level has its story, how do they relate?" He seemed to like that, and a few months later, emboldened by this encounter, I invited him to a small meeting I was organizing on memory in Moorea. He instantly accepted. Crick was always annoyed and impatient when hearing about the status quo of any topic. He liked a good experiment, but he always sought to know what a particular observation meant in a larger context. At the meeting, he was no different. He always shook things up. He seemed to be just the person needed to push consciousness studies forward beyond the well-honed classic positions.

Crick began by teaching himself neuroanatomy and reading extensively on neurophysiology and psychophysics. In 1979, a couple of years into this endeavor, he was asked to write an article for an issue of *Scientific American* dedicated to the latest in brain research. His assignment: "to make some general comments on how the subject strikes a relative outsider." Crick noted that he was not happy with behaviorists and functionalists treating the brain as a black box. After all, it was the very workings inside the black box that were in question. "The difficulty with the black-box approach is that unless the box is inherently very simple a stage is soon reached where several rival theories all explain the observed results equally well."[31] And nobody thought that the black box was simple.

Crick also noted that brain scientists were too isolated in their particular subdisciplines. They needed to be less scientifically provincial and more scientifically cosmopolitan by engaging in more interdisciplinary cross talk. Psychologists needed to understand the structure and function of the brain, just as anatomists needed to know about psychology and physiology. Of course, his broad assessments were at the expense of dozens if not hundreds of cognitive scientists[32] and budding young cognitive neuroscientists like yours truly,[33] who had already begun to toil with the problem of consciousness. Yet it was Crick, with his unique and special status, who yanked the field into realizing that studying the physical basis of consciousness was a crucial task.

Everyone needed a dash of neuropsychology, a bit of physics and chemistry, and Crick thought the new field of communication theory held promise as a theoretical tool, so get a handle on that, too. An overarching theory was only going to happen when all aspects and levels of the brain's churnings and of human behavior were accounted for. If you were only familiar with one aspect, you didn't have a chance at any sort of all-encompassing explanation.

One of Crick's suggestions was particularly difficult. He proposed that we needed to change people's mind-set about the accuracy of their own introspection, because "we are deceived at every level by our introspection."[34] One example he used of that deception is the blind spot that we have in each eye. Crick also chastised current philosophers—though we can assume Dennett is not on the list—for ignoring such phenomena:

> Not everyone realizes he has a blind spot, although it is easy to
> demonstrate. What is remarkable is that we do not see a hole in our
> visual field. The reason is partly that we have no means of detecting
> the edges of the hole and partly that our brain fills in the hole with
> visual information borrowed from the immediate neighborhood.
> Our capacity for deceiving ourselves about the operation of our
> brain is almost limitless, mainly because what we can report is only

a minute fraction of what goes on in our head. This is why much of philosophy has been barren for more than 2,000 years and is likely to remain so until philosophers learn to understand the language of information processing.

This is not to say, however, that the study of our mental processes by introspection should be totally abandoned, as the behaviorists have tried to do. To do so would be to discard one of the most significant attributes of what we are trying to study. The fact remains that the evidence of introspection should never be accepted at face value. It should be explained in terms other than just its own.[35]

Crick concluded:

The higher nervous system appears to be an exceedingly cunning combination of precision wiring and associative nets. . . . The net is broken down into many small subnets, some in parallel, others arranged more serially. Moreover, the parcellation into subnets reflects both the structure of the world, external and internal, and our relation to it.[36]

Crick was a theorist at heart, with a particular genius for assimilating ideas and experimental results from a wide range of disciplines, churning them up together, and then formulating new theories and new experiments. He clearly stated the deep problems involved in trying to understand conscious experience. He had the invaluable skill of doing what William James suggested: "The art of being wise is the art of knowing what to overlook." And wise he was.

Crick soon teamed up with the clever, smart, and endlessly energetic computational neuroscientist Christof Koch at Caltech. They decided to tackle consciousness by setting to work on the visual system of mammals—a topic that had already generated a plethora of experimental data. Their aim was to learn as much as possible about the first processing steps performed on visual information that is received by the visual cortex. Their ultimate goal was to discover the neural correlates of consciousness (NCC): the minimal set of neuronal events and mechanisms jointly sufficient for a specific conscious percept.[37] Koch explains: "*There must be an explicit correspondence between any mental event and its neuronal correlates.* Another way of stating this is that any change in a subjective state must be associated with a change in a neuronal state. Note that the converse need not necessarily be true; two different neuronal states of the brain may be mentally indistinguishable."[38] This all sounds eminently reasonable and straightforward, a quality in short supply in research on consciousness.

To begin their quest, Crick and Koch made two assumptions about consciousness. One was that at any one moment some active neuronal processes correlate with consciousness, while others do not. They ask: What are the differences between them? The second assumption they called "tentative": that "all the different aspects of consciousness (smell, pain, vision, self-consciousness . . . and so on) employ one or perhaps a few common mechanisms."[39] If they understood one aspect, they would be on the road to understanding them all. They decided to shelve some discussions in order to avoid wasting time quibbling over them. Skirting the mind/body stalemate, they contended that in order to examine consciousness scientifically, since everyone had a rough idea of what was meant by consciousness, they didn't need to define it, and thus would avoid the dangers of a premature definition.

Since they were being vague about its definition, Crick and Koch decided to be consistently vague about its function and set aside the question of what consciousness is for. They also chose to assume that some species of higher mammals possess some features of consciousness but not necessarily all. Thus, one may have key features of consciousness without having language. And while lower animals may have some degree of consciousness, they didn't want to deal with that question at the moment. They assumed that self-consciousness was a self-referential aspect of consciousness and put that aside, too. They also put volition and intentionality aside for the moment, along with hypnotic states and dreaming. Finally, they put qualia aside—the subjective character of experience, the feeling of "red"—believing that once they figured out how one saw red, then perhaps a plausible case could be made that our reds are indeed all the same.

Crick and Koch both acknowledged that the NCC would not solve the mystery of consciousness. What identifying the neural correlates of conscious versus nonconscious processing would do for the empirical study of consciousness would be to supply constraints on the specifications of neurobiologically plausible models. The hope is that elucidating the NCC might provide a breakthrough for the theory of consciousness similar to what the structure of DNA did for genetic transmisson. Understanding the architecture of the DNA molecule and making a 3-D model provided clues to how the molecule broke apart and replicated itself, which correlated very well with Mendelian inheritance. The first definitive NCCs discovered will be early steps toward a theory of consciousness, but they in themselves will not provide explanations of the links between neural activity and consciousness. That is what models do—and soon, a fresh crop began to appear.

Crick opened the floodgates: it was okay to study consciousness again. In those previous two decades, a foundation had been laid, built of stacks of empirical data about brain mechanisms. An empirical attack was launched, aided by an ever-increasing arsenal of new methods, which now range from not just recording but controlling the firing of single neurons (a goal Crick longed for that has since been achieved through optogenetics), to various types of brain imaging, to all the data crunching that computers can provide. Those who heeded Crick's warning that "introspection should never be accepted at face value. It should be explained in terms other than just its own" found an embarrassment of riches in the brain's nonconscious processing. Neurobiological models that attempted to explain the links between neural activity and consciousness, using computational, informational, and neurodynamic elements, began to pop up like mischievous ideas in a rascally child. The models vary according to the level of abstraction they address—

something we'll tackle in chapter 5—and while some have shared features, none explains all aspects of consciousness, and none has yet won general approval.

In the upcoming chapters, I plan to lay out a new idea and framework for thinking about the problem of consciousness. I do it humbly and nervously; trying to add to the story previously generated by this pantheon of thinkers and scientists is daunting, to say the least. Still, today we have at our fingertips a vast amount of rapidly accruing new information, and with a little luck, it affords new perspective on how the brain does its magic. The ideas of Descartes and other past thinkers that the mind is somehow floating atop the brain, and the ideas of the new mechanists that consciousness is a monolithic thing generated by a single mechanism or network, are simply wrong. I will argue that consciousness is not a thing. "Consciousness" is the word we use to describe the subjective feeling of a number of instincts and/or memories playing out in time in an organism. That is why "consciousness" is a proxy word for how a complex living organism operates. And, to understand how complex organisms work, we need to know how brains' parts are organized to deliver conscious experience as we know it. That is up next.

PART II:

THE

PHYSICAL

SYSTEM

4.

MAKING BRAINS ONE
MODULE AT A TIME

"Now, here, you see, it takes all the running you can do,
to keep in the same place. If you want to get somewhere
else, you must run at least twice as fast as that!"
—The Red Queen, in Lewis Carroll's *Through the Looking-Glass*

OUR BRAINS LOOK as wonky as Frank Gehry's Guggenheim Museum in Bilbao, Spain, but as Gehry is apt to point out, the museum doesn't leak. It works! Gehry is an architectural genius who expanded our imagination of physical structures that perform useful functions. Our brain also has a physical structure that performs useful functions. There is method in the madness of that wonky-looking structure, bits of which we understand, most of which we don't. Despite centuries of research, nobody fully understands how the convoluted mesh of biological tissue inside our heads produces the experiences of our everyday life. Gazillions of electrical, chemical, and hormonal processes occur in our brain every moment, yet we experience everything as a smoothly running unified whole. How can this be? Indeed, what is the organization of our brain that generates conscious unity?

Everything has an underlying structure; physicists take this truth down to the quantum level (which we will discuss in chapter 7). We are constantly taking things apart to see what makes them tick. Things are made up of parts, and so are bodies and brains. In that sense one could say we are built out of modules, which is to say, components that interact to produce the whole functioning entity we are examining. We need to know the parts and not only how they all mesh together but also how they interact.

There is little doubt that in some way the parts of our brain work collectively to produce our mental states and behaviors. On the surface, it seems logical to think that our brain functions as a global unit to produce a single conscious experience. Even the Nobel laureate Charles Sherrington, writing in the early 1900s, described the brain as an "enchanted loom,"[1] suggesting that the nervous system works coherently to create the mystical mind. Yet neurologists at the time would have suggested to him to go on medical rounds. Their clinics were full of patients whose brain injuries told a different story.

Paradoxically, while all of us *feel* like a single undivided entity (a fact that seems to provide intuitive evidence for Sherrington's loom), considerable evidence suggests that the brain does *not* operate in a holistic fashion. Instead, our undivided consciousness is actually produced by thousands of relatively independent processing units, or, more simply, *modules*. Modules are specialized and frequently localized networks of neurons that serve a specific function.

The neuroscientist, physicist, and philosopher Donald MacKay once commented that it is easier to understand how something works when it is not working properly. From work in the physical sciences, he knew that engineers could more quickly figure out how something, such as a television, worked if the picture was flickering than when it was running smoothly.[2] Similarly, studying broken brains helps us understand better how unbroken ones work.

The most compelling evidence for a modular brain architecture arises from the study of patients who have suffered a brain lesion. When damage occurs to localized areas of the brain, some cognitive abilities will be impaired because the network of neurons responsible for that ability no longer functions, while others remain intact, tootling along, performing flawlessly. What is so intriguing about the brain-altered patients is that no matter what their abnormality, they all seem perfectly conscious. If conscious experience depended on the smooth operation of the entire brain, that shouldn't be what happens. Since this fact—that modules are everywhere—is so central to my thesis, it's important that we understand how modular the brain truly is.

Missing Modules but Functioning Brains

Take a lobe, any lobe in the brain, and consider people who have suffered a stroke. People with a right parietal lobe injury, for example, will commonly suffer from a syndrome called *spatial hemi-neglect*. Depending on the size and location of the lesion, patients with hemi-neglect may behave as if part or all of the left side of their world, which may include the left side of their body, does not exist! This could include not eating off the left side of their plate, not shaving or putting makeup on the left side of their face, not drawing the left side of a clock, not reading the left pages of a book, and not acknowledging anything or anyone in the left half of the room. Some will deny that their left arm and leg are theirs and will not use them when trying to get out of bed, even though they are not paralyzed. Some patients will even neglect the left side of space in their imagination and memories.[3] That the deficits vary according to the size and location of the lesion suggests that damage that disrupts specific neural circuits results in impairments in different component processes. Mapping the functional neuroanatomy of these lesions has provided strong evidence for this suggestion.[4]

Now, here is the kicker: while hemi-neglect can occur when there is actual loss of sensation or motor systems, a version of it can also occur when all sensory systems and motor systems are in good working order—a syndrome known as extinction. In this case each half brain seems to work just fine alone, but it begins to fail when required to work at the same time as the other half. Yet information in the neglected field can be used at a nonconscious level![5] That means the information is there, but the patient isn't conscious it is there. Here is how it works. If patients with left hemi-neglect are shown visual stimuli in both their right and left visual fields at the same time, they report seeing only the stimulus on the right. If, however, they are shown only the left visual stimulus, hitting the same exact place on the retina as previously, the left stimulus is perceived normally. In other words, if there is no competition from the normal side, then the neglected side will be noticed and pop into conscious awareness! What is strangest of all is that these patients will deny that there is anything wrong; they are not conscious of the loss of these circuits and their resulting problems.

It appears, then, that their autobiographical self must be derived only from what they are conscious of. And what they are conscious of is dependent on two things. First, they are not conscious of circuits that are not working. It is as if the circuits never existed and consciousness for what the circuits did disappears with the circuit. The second thing is that some sort of competitive processing is happening. Some circuits' processing makes it into consciousness while others' does not. In short, conscious experience seems tied to processing that is exceedingly local, which produces a specific capacity, and that processing can also be outcompeted by the processing of other modules and never make it to consciousness. This has astounding implications.

While some patients are not conscious of parts of their body that are actually there, my all-time favorite clinical disorder is the "third man" phenomenon,* in which a person feels the presence of another who actually

*Ernest Shackleton first described the feeling of a presence. He and his two companions were in a state of utter exhaustion and physical deprivation after successfully crossing 680 miles of the world's roughest seas in a leaky lifeboat with scant food and water. They were on the final leg of an epic mission to obtain help for his stranded crew left on an island off the coast of Antarctica: crossing two unmapped, snow-covered mountain chains on South Georgia Island, with only an ice ax and fifteen yards of rope, as quickly as possible. While on this trek, Shackleton described the feeling that they were being accompanied by a fourth man. Later, T. S. Eliot made a reference to this phenomenon of feeling a presence in his poem "The Waste Land," but called it "the third man," and this is what has stuck (J. Geiger, *The Third Man Factor: Surviving the Impossible* [New York: Weinstein Books, 2009]).

is not there! Known as a "feeling of a presence" (FoP), it is the sensation that someone else is present in a specific spatial location, often just over the shoulder. It is so strong that people will continue to turn their head to glimpse or offer food to the presence. This is not the same as walking down a dark alley and getting creeped out by imagining someone following you. This presence pops up unexpectedly. It is actually a common phenomenon among alpinists and others suffering intense physical exhaustion in extreme conditions.

In his book *The Naked Mountain*,[6] Reinhold Messner, widely considered to be the greatest mountaineer of all time (he was the first to solo-climb Mount Everest and, incidentally, never uses supplemental oxygen), described what happened in 1970 while he was making his first major Himalayan ascent, of Nanga Parbat, with his brother Günther: "Suddenly there was a third climber next to me. He was descending with us, keeping a regular distance a little to my right and a few steps away from me, just out of my field of vision. I could not see the figure and still maintain my concentration but I was certain there was someone there. I could sense his presence; I needed no proof." You don't have to be an exhausted alpinist, however, to experience such a presence. Nearly half of widows and widowers have felt the presence of their deceased spouses.[7] For some, such phenomena are the starting point for tales of apparitions, ghosts, and divine intervention.

Not so, claims the Swiss neurologist and neurophysiologist Olaf Blanke, who came across the phenomenon unexpectedly. He had triggered it with electrical stimulation to the temporal parietal cortex of a patient's brain while trying to locate the focus of a seizure.[8] He has also studied a bevy of patients who complain of an FoP. He found that lesions in the frontoparietal area are specifically associated with the phenomenon and are on the opposite side of the body from the presence.[9] This location suggested to him that disturbances in sensorimotor processing and multisensory integration may be responsible. While we are conscious of our location in space, we are unaware of the multitude of processes (vision, sound, touch, proprioception, motor movement, etc.) that, when normally integrated, properly locate us there. If there is a disorder in the processing, errors can occur and our brains can misinterpret our location. Blanke and his colleagues have found that one such error manifests itself as an FoP. Recently, they cleverly induced the FoP in healthy subjects by disordering their sensory processing with the help of a robotic arm.[10]

When we make a movement, we expect its consequence to occur at a specific time and location in space. You scratch your back, you expect to feel a sensation simultaneously on your back. When the sensation is spatially and temporally matched as expected, your brain interprets the sensation as self-generated. If there is a mismatch, if the signals are spatially and temporally incompatible with self-touch, you rule it as being done by another agent. Now picture yourself blindfolded, arms extended in front of you, with your fingertip in the thimble-like slot of a "master robot" that sends signals to a robotic arm behind your back. Your finger movements control the robotic arm's movement, which strokes your back as you move your finger. In some trials your finger feels resistance that matches the force with which it is pushing, and in others the resistance is loosey-goosey, not properly correlated to what you are doing. If the sensation on your back is synchronous with your movement, even with your arms extended in front of you, your brain creates an illusion: you will feel as if your body has drifted forward and that you are touching your own back with your finger. If, however, the touch sensation is not synchronous, if it comes a tick late, your brain cooks up something different. Your self-location drifts in the opposite direction, backward away from your fingertip; you feel as if something other than you is touching your back. If, in addition, you also felt no resistance in the fingertip while controlling the arm, this asynchronous touch produces a feeling that another person is behind you touching your back! Blanke, using well-controlled bodily stimulations, demonstrated that sensorimotor conflicts (that is, signals that are spatially and temporally incompatible with physical self-touch) are sufficient to induce the FoP in healthy volunteers. These conflicts were produced by manipulating different localized neural networks—modules.

If the brain worked as a consolidated "enchanted loom," then removing portions of the brain or stimulating erroneous processing in some circuits would either shut down the system entirely or cause dysfunction across all cognitive realms. In reality, many people can live relatively normal lives even if portions of their brain are missing or damaged. When people have damage to localized brain areas, there almost always appears to be impairment in some, but not all, cognitive domains. For example, a well-developed cognitive domain in humans is language. The language center in most people is housed in the left hemisphere. Two very distinct brain regions within the language center include Broca's area and Wernicke's area.

Broca's area contributes to speech production, whereas Wernicke's area deals with comprehension or understanding of written and spoken language and helps organize our words and sentences in an understandable way. Specifically, Broca's area is involved with word articulation, coordinating the muscles in our lips, mouth, and tongue to accurately pronounce words, while Wernicke's area organizes our words in a comprehensible order before we even speak. People with damage to Broca's area have difficulty speaking: speech is effortful and comes in bursts, but the words they manage to announce are in a comprehensible order (e.g., "Brains . . . modu . . . lar"), though it may lack proper grammar. Broca patients are aware of their errors and are quickly frustrated. Conversely, people with damage to Wernicke's area primarily have a comprehension disorder. They have speech with normal prosody and correct grammar, but what they say makes no sense. This shows us that each of these areas has a different and specific job; if that area is damaged, then it can no longer perform the job properly. This unambiguously demonstrates that there is hyperspecific modularity in the brain.

Why did modularity evolve in brains? I once heard the CEO of Coca-Cola describe the logic of the company's corporate organization. As the company grew, the executives realized that having a central plant that made all of their product and then shipped it out to the world was crazy, inefficient, and costly. The shipping, the packaging costs, the travel costs of holding management meetings in "corporate headquarters," and on and on made no sense. Clearly, they should divide the world into regions, build plants in each of those regions, and distribute their product locally. Central planning was out, local control was in. Same for the brain: cheaper and more efficient.

Evolving a Bigger Brain

Historically, it was assumed that animals with brains bigger than expected for their body size had greater intelligence and abilities. Humans were thought to have overly large brains, proportionally, and this accounted for our diverse abilities and intelligence. This theory has always had a problem, however. Neanderthals actually had bigger brains than we do, yet they didn't make the competitive cut when *Homo sapiens* arrived on the scene. From my own research there appears another thorny problem: after split-brain surgery, the stand-alone left hemisphere (one-half of the brain) is nearly as intelligent as the intact whole brain. Bigger isn't necessarily better. What is going on?

Suzana Herculano-Houzel and her coworkers, armed with a new technique to count neuron and non-neuron cell numbers in human brains, compared them across species. They found that rumors of our big brains were greatly exaggerated! The human brain is not out of whack size-wise but is a proportionately scaled-up primate brain. Although the human brain is much larger and has many more neurons, the *ratio* of neuron number to brain size is the same for chimps and humans.[11] Another rousing finding was that the often-cited *ratio* of glial cells to neurons of 10:1, albeit tossed around with no references ever noted, was completely off the mark. In fact, the human brain holds no more than 50 percent glial cells, just as other primates' brains do. Busting another myth, Herculano-Houzel suggests that this overestimate of the glial-to-neuron ratio of 10:1 may have been the basis for the false notion that we use only 10 percent of our brains![12]

Yet compared to the brains of other mammals, the human brain has two advantages. First, it is built according to the very economical, space-saving scaling rules that apply to other primates, and among those economically built primate brains, it is the biggest and thus contains the most neurons. Brain size, however, cannot be used willy-nilly as a proxy for neuron number when comparing other species with primates. For instance, in rodents, comparing mice and rats, the rat brain is bigger but not solely because it has more neurons. For the rat, when the number of neurons increased, so did their size. Thus, a single rat neuron takes up more volume than a single mouse neuron: it's like the size difference between capellini and spaghetti noodles. In primates, when comparing monkeys to humans, however, as neuron numbers increase, the size of the neuron stays the same. The result is that a bigger primate brain has a greater net increase in the number of neurons per volume than a bigger rodent brain. If we took a rat brain and increased its volume to be equal to that of a human brain, the rat would have only 1/7 the number of neurons as a human brain, simply because each of his neurons would take up more space. Increasing brain size is a tricky business, and it looks as if different orders (Primata, Rodentia, and the like) follow different rules when it comes to scaling up.

And this brings us back to modules. If, as human brains increased in neuron number, every neuron were connected to every other one, the number of axons (the cable part of each neuron) would increase exponentially. Our brains would be gigantic—in fact, they would be twenty kilometers in diameter[13] and require so much energy that even if we were force-fed like a Toulouse goose, we would still not be able to run the thing.[14] As it is, our brains represent about 2 percent of our entire body weight and suck down about 20 percent of our energy. Brains use so much energy because they are powerful electrical systems that are constantly active, like an air conditioner in July, in Phoenix. Another problem would be that the axons would be so long that the processing speeds would take a nosedive.

The neuroscientist Georg Striedter studies how and why differences occurred in brain evolution in different species. He suggests that there are certain laws that govern connectivity as brains increase in size.[15] First of all, the number of neurons that an average neuron connects to does not change with increasing size. Instead, the absolute number of neuronal connections stays the same, with the result that the increase in brain size would be more manageable in terms of energy requirements and space. That means, however, that there is decreased connectivity overall as brains enlarge. Decreased connectivity means more independent processing.

The second law is that the connection lengths are minimized. This results in most neurons being hooked up to neighboring neurons. Short connections take less energy, less space, and less time for signaling, producing efficient communication between these localized neurons. Thus, as brains enlarge, wiring reorganization ensues, and their structural architecture changes. The resulting structural architecture is one of clusters of well-connected localized neurons, or "communities."

This type of organization allows these separate clusters to independently specialize in performing a certain function: a module is born! While most neurons in a module sport intramodular connections, a sparse few have short connections to neurons in neighboring modules, allowing the formation of a neural circuit. A neural circuit is formed when a module receives information, modifies it, and transmits it to another module for further modification. Thus, while most modules are sparsely connected with other modules, the wiring does allow neighboring modules to form clusters for more complex processing. We will learn more about this in the next chapter when we discuss layered architecture.

Some modules are hierarchically arranged, such that they are made up of submodules, which themselves are made up of sub-submodules.[16] Yet multiple modules running independently create a need for some efficient communication and coordination between them. This brings us to the third wiring requirement: Not all the connections are minimized; some long connections are maintained and serve as "shortcuts" between distant sites.

The overall architecture that these wiring laws produce is known as "small-world" architecture. This type of architecture is famous for its ability to host a high degree of modularity, yet few steps are needed to connect any two randomly selected modules. Small-world architecture is found in many complex systems, such as the western U.S. power grid and social networks. Multiple studies have borne out the notion that the brain is organized into clusters or modules of functionally interconnected regions.[17]

Advantages of a Modular Brain

Looking at this design, we can see that there are many reasons why a modular brain is superior to a globally functioning brain. First of all, a modular brain cuts down on energy costs. Since it is divided into units, only regions within a given module need to be active to complete specific assignments. If you used your entire brain for every action, your brain's electric bill would go through the cranium. It is the same deal in Phoenix in the summer. If you only have the AC running in the bedroom at night, it is cheaper than if it is cooling the entire house. Despite saving energy through modularity, is the brain really that energy efficient if a fifth of your diet is dedicated to powering the thing?

It turns out the brain is fairly efficient despite being an energy hog. Neurons transmit electrical impulses through axons and dendrites, the brain's "wires." Although neural wiring functions much differently than the wiring in modern electrical devices, the basic idea is the same: electrical current transfers information from one place to another, and this requires energy. The farther an electrical current travels, the more energy it consumes; and the thicker the axon, the more resistance encountered and thus the more energy needed to overcome it. By working in local modules, the brain saves energy by operating over short distances, with thin wires, with short conduction times for information traveling between those modules. Additionally, given the dynamics of neural systems, a 60 percent wire fraction (the proportion of gray matter that is made up of axons) is what is predicted if wire length and thus conduction delays are minimized. Many brain structures are composed of wiring systems that are near this optimal value.[18] If, instead, brains functioned as a global unit, then each brain region would have roughly equal amounts of wiring for short- and long-distance communication, and longer distances mean more "wire" and thus more "cost." A modular brain cuts costs by keeping the wiring to a lower 3-to-5 ratio (the 60 percent wire fraction), thereby limiting the number of distantly communicating electrical transmissions. Overall, the brain appears to maximize energy efficiency by operating in modules.

Modular brains are also functionally efficient, because multiple modules can process specialized information simultaneously. It is much easier to walk, talk, and chew gum at the same time if many modular systems are working independently, rather than a single system attempting to coordinate all the actions. Plus, if the brain behaved as a single unit, then it would need to be a "jack-of-all-trades" to adequately perform all of our daily duties. It is more efficient to have specialized "master" modules perform specific tasks. Specialization is ubiquitous in complex systems. For instance, economies thrive when the best farmers farm, the best educators educate, and the best managers manage. Bad managers can sink a business, bad farmers can go bust, and bad teachers—well, we have all suffered at least one of those and know the consequences. When people apply themselves and focus on specific jobs without being concerned about all jobs necessary to keep an economy running, they become experts. Experts are more efficient producers. When experts work simultaneously, there is greater economic output than there would be if everyone attempted to do a little bit of everything. Thus, it seems reasonable to think that our brains evolved in a modular way to efficiently process multiple types of information concurrently.

Perhaps most important, a modular brain also allows faster adaptation or evolution of the system to a changing environment: because one module can be changed or duplicated independently of the rest, there is no risk of changing or losing other, already well-adapted modules in the process. Thus, further evolution of one part does not threaten well-functioning aspects of the system.

Even if we take evolution out of the equation, brain modularity is helpful in acquiring new skills. Researchers have found that the architecture of particular networks changes over the course of learning a motor skill.[19] Although many skills take considerable time to perfect, we are able to learn

new skills through experience. If the entire brain changed the way it functioned whenever we acquired a new skill, we would lose our expertise in old
skills. The perks of brain modularity are that it saves energy when resources
are scarce, allows for specialized parallel cognitive processing when time
is limited, makes it easier to alter functionality when new survival pressures
arise, and allows us to learn a variety of new skills. When one stops to think
about it, how could the brain possibly be organized any other way?

Going Modular

Human brains are neither the only modular brains nor the only modular biological systems. Worm brains, fly brains, and cat brains are modular, as are
vascular networks, protein-protein interaction networks, gene regulation
networks, metabolic networks, and even human social networks.[20] How did
this modularity evolve? What selection pressures produce a modular system?
This was the question that puzzled a trio of computer scientists who, after
mulling it over, decided to test Striedter's hypothesis that modularity is the
by-product of pressure to minimize connection costs.[21]

Construction costs in a network include the costs of manufacturing the
connections and maintaining them, and the energy it takes to transmit along
them, as well as the cost of signal delays. The longer the connections and the
more there are, the more expensive the network is to build and maintain.[22]
Also, adding more connections or length to a signaling pathway could delay
critical response times—not good for survival in a competitive environment
when a predator starts salivating at the sight of you, bares its fangs, and flexes
its claws.

The computer scientists Jeff Clune, Jean-Baptiste Mouret, and Hod
Lipson did what computer scientists do: they designed computer simulations.[23] They used well-studied networks that had sensory inputs and produced outputs. What those outputs were determined how well the network
performed when faced with environmental problems. They simulated twenty-
five thousand generations of evolution, programming in a direct selection

pressure to either maximize performance alone or maximize performance *and* minimize connection costs. And voilà! Once wiring-cost-minimization was added, in both changing and unchanging environments, modules immediately began to appear, whereas without the stipulation of minimizing costs, they didn't. And when the three looked at the highest-performing networks that evolved, those networks were modular. Among that group, they found that the lower the costs were, the greater the modularity that resulted. These networks also evolved much quicker—in markedly fewer generations—whether in stable or changing environments. These simulation experiments provide strong evidence that selection pressures to maximize network performance and minimize connection costs will yield networks that are significantly more modular and more evolvable.

So now we know that modular systems have many advantages, but how do they do it? How do thousands of independent localized modules work together to coordinate our thoughts and behaviors and, ultimately, produce our conscious experience?

Modular Connections

Although modules are highly intraconnected to compute specialized functions, we have learned that they are also loosely connected to other modules. Some communication between modules is vital for coordinating complex behaviors. For instance, Broca's and Wernicke's areas have their own specialized functions for language, but they also must converse with each other. Wernicke's area needs to organize phonemes and words into coherent sentences in order for Broca's area to guide your lips, mouth, and tongue to produce the correct sound sequence. These language areas are densely connected via the arcuate fasciculus, a bundle of nerve fibers that runs between them like a highway. Your brain minimizes these costly, large communication networks by reducing connections between modules that contribute to different types of cognitive functions. There is no need to activate Broca's and Wernicke's areas when you're smelling a rose, unless you launch into a sonnet on its beauty or a rant about hybridizers selecting for form over fragrance. Brain modules communicate with one another, but there are disproportionately more connections between modules that perform related cognitive processes and many fewer connections between modules involved in dissimilar processes.

Animal and Human Brains: What's the Difference?

Even when employing strikingly different methods and data analysis techniques, most studies provide evidence that modules, in both structural and functional brain networks, exist across all species and share many of the same properties.[24] It is worthwhile to take a moment to understand the difference between a structural and a functional network. "Structure" refers simply to the physical anatomy of a network: how many neurons, how they are arranged, their shape, and so forth. A functional network performs a certain function; it may have to do with speaking language, or it may have to do with understanding language. Importantly, the structure of a network does not reveal its function, or vice versa. It may lend clues, but that is it. For example, you can look at a tree and see its structure, but that tells you nothing about the function of leaves. Studies involving animals, ranging from invertebrates to mammals, have revealed that their neural modules are also highly intraconnected and spatially close to one another to reduce energy consumption. Interestingly, the neuronal network of the transparent nematode *Caenorhabditis elegans* (an organism possessing a few hundred highly studied neurons) functions in modules as well, despite its being one of the tiniest creatures with a neural system.[25] Across species, modularity is efficient and necessary for organisms to effectively function and evolve in a competitive environment.

It is natural to assume that if modular brains are present in animals and humans, they must share similar cognitive aspects, including consciousness. Unfortunately, even though Thomas Nagel would love it, technology today does not permit us to truly understand how different organisms experience the world. Often it is even difficult for us to understand our own perception of the world. The best we can do to empirically understand the experience of others, both animals and people, is to use behavioral and brain-activity measurements.

It is not surprising that we associate conscious experience with our human complex cognitive skills. We jump to the conclusion that an animal, in order to be conscious, must have those same types of skills. We freely map the capacity to experience consciousness onto any number of things, from puppets to robots to, in my case, a 1949 Plymouth coupe.

One way researchers have looked for clues of early conscious states in other animals is to look for signs of tool use. Using tools is one behavior that is considered to indicate complex cognition. It turns out signs are all over the animal kingdom. Corvid birds, for example (in the Corvidae family: crows, ravens, jays, magpies, rooks, nutcrackers), develop tools to get food from hard-to-reach places in a manner similar to the way chimpanzees manufacture and use tools.[26] Japanese crows in Sendai City use cars to crush nuts: they drop them onto pedestrian crossings and not only wait until they are crushed but also wait until the light turns red before retrieving them. New Caledonian crows are the whiz kids, making two types of tools, which they use in different ways for different jobs. They carry them when they go foraging, the way a fisherman carries his pole. They also solve "meta-tool" problems, in which they have to use one tool to get a second tool that is needed to retrieve food.[27] Crows from different areas have different tool designs, suggesting that they show cultural variation and transmission.[28] But basic stick-tool skills can be developed by hand-raised crows without social learning.[29] While this most likely means crows are conscious in the sense that they are alive, alert, and experiencing the moment, does it indicate they are aware of their skills? They surely have some specialized modules other birds do not have, but does that make them self-aware? The many studies of their behavior, skills, and learning don't venture to tackle that question. What about chimps?

Chimpanzees in the wild have long been observed using tools, primarily sticks to scoop up ants and honey, and leaves to scoop up water. Chimps from different geographical locations also use different tools for different purposes—here, too, suggesting cultural variation and social transmission of tool use. Yet, once a tool-use behavior has been learned by a chimp, it becomes a habit, and chimps don't upgrade to an improved technique if one is discovered and used by a few members of their group.[30] On the other hand, chimps that have been living with humans have been observed to solve puzzles and find solutions to complex problems. For example, chimpanzees that had seen a banana hanging from the ceiling out of reach stacked crates on top of one another to create a makeshift ladder to retrieve the banana.[31] While the list of chimp tricks is long and dazzling, does this make them conscious beings in the same sense that humans are conscious? This is probably an ill-posed question. Perhaps the question should be "Does our conscious experience hold similar contents to that of a chimp?"

Chipping away at what all of these animal studies mean, many studies have compared chimpanzee thinking to infant thinking. In a simple hidden-object pointing task, where a desired object is placed out of view and the experimenter points to where it is, chimps are bewildered, whereas human infants succeed at fourteen months of age.[32] At the same time, if a chimpanzee or a child observes a behavior, they both can mimic the behavior even if they have never performed such an act before. While children imitate all actions that they are shown to attain a reward goal, even the superfluous ones, chimps imitate only the necessary ones. It has been suggested that this indicates that children are compulsive imitators, whereas chimps imitate to attain a goal. If a chimpanzee is not presented with a reward (or punishment), however, then the learned behavior is generally not repeated. In contrast, infants will mimic behaviors regardless of whether there is a reward or punishment, suggesting that human infants have a propensity to learn new behaviors for the sake of learning alone.[33] This would constitute a huge difference between humans and the rest of the animal kingdom. Still, it seems that the chimps have more going on than the corvids. Does their added cranial hardware enable conscious experience or simply change its contents?

Humans have an ability to learn and solve abstract problems that exceeds the capacity of other animals. People have invented technologies that are more sophisticated and provide much more utility than any tool an animal has created. Engineers and scientists have developed computers, airplanes, skyscrapers, rockets that take us to the moon . . . you name it. We only need a small portion of the population to be inventive, however. Through imitation and learning, useful items and discoveries spread like wildfire through the population and become part of our everyday lives. As the gifted psychologist David Premack has pointed out, humans have a "select few" who can develop great technologies, such as controlling fire, the wheel, agriculture, electricity, cell phones, the Internet, and bacon- and cheese-stuffed potato skins. No other living species have *any* members that achieve such great feats.[34] Is all of this extra capacity to learn, problem-solve, and invent what allows us to be conscious? There has to be hardware, which is to say special modules, that allows for these capacities. Are they the key to our understanding?

I find the whole line of thinking that there is a magic potion that produces human consciousness misguided. Seeing all the marvelous things a chimp can do springs our minds into action, and we confer on them a huge special status. We grant them entry into our consciousness club, and we are happy to do so. But it took the person who discovered and articulated the mental life of chimpanzees to ask the question: What do they think about it all? We have a theory about chimps, but do they have a theory about us?

Premack, along with his student Guy Woodruff, was the first to test whether other animals have a "theory of mind."[35] Possessing a theory of mind means that an individual ascribes mental states, such as purpose, intention, knowledge, beliefs, doubts, pretending, liking, and so forth, to himself and to others. Premack and Woodruff, who coined the term, called it a "theory" because such states in others are not directly observable; they are inferred. Humans assume that others have minds and that their mental states drive their actions. Almost forty years after the idea was proposed, the dust still hasn't settled, but it appears that while some animals do possess some degree of theory of mind, none have it to the extent that humans do. Josep

Call, Michael Tomasello, and their colleagues have spent many years whittling away at this question. Chimps understand the goals and intentions of others, and the perceptions and the knowledge of others, to some extent, but despite many attempts to prove otherwise, it appeared that chimpanzees do not understand that others may have false beliefs,[36] a test that two-and-a-half-year-old children pass.[37] Just recently, however, Call and Tomasello, along with Christopher Krupenye, have found evidence that suggests three species of great apes do have some implicit understanding that others have false beliefs, but they have not yet been shown to make explicit behavior choices based on an understanding of false beliefs.[38] How close the apes' theory-of-mind ability is to that of humans remains to be seen.

Dogs have recently started sharing the limelight in the animal IQ contests in which sociability is concerned. Chaser, the retired psychology professor John Pilley's famous border collie, knows over a thousand words, understands syntax, and makes inferences about what novel words mean.[39] If he is told to bring the "dax" (a word he has never heard before), he will look through his substantial pile of toys and bring the one he has never seen before. Dogs can also make inferences about hidden food and other hidden objects based on social cues, such as human pointing (something that chimps do not do). Michael Tomasello suggests that this involves understanding two levels of intention: what and why. First, the dog must understand that the pointer intends for her to attend to what is being pointed at, and, second, the dog must figure out why she is supposed to: Is the person offering helpful information on where something is located, or does the person want the object for himself?[40] While a chimp often follows the pointing gesture, chimps don't understand that the food is hidden there: they don't seem to figure out the second level of intention, the why. In the past twelve years, the impressive ability of dogs to use communicative cues made by humans has sparked the interest of some researchers studying theory of mind, and while there are early indications that dogs do possess it to some extent,[41] much more research needs to be done.

While dog lovers are thrilled with these findings, it should be remembered that dogs don't show any special flexibility in nonsocial domains. They are slaves to special interests—I mean special capacities. When presented with nonsocial cues, such as food hidden under a tilted-up board versus one lying flat, they can't solve the problem (an easy problem for a chimp to solve), nor do they understand that they should preferentially grab a string with food attached to it rather than one that is not attached to the food, again something that a chimp clues in to right away.[42] The differing cognitive abilities of dogs suggest that they possess specific yet different modules, which evolved in response to different environmental pressures. The contents of their conscious experience are different from ours and different from chimps', though some, no doubt, are shared.

Overall, it looks as if trying to nail down a cognitive prerequisite for conscious experience is a mug's game. A little of this and a little of that doesn't quite capture what the brain must do to engender conscious experience. The brain is not going to give up this trick easily, if indeed it is a trick. Recall that we do not consciously experience the blind spot in our visual field, even though it is there. Our visual system is performing a consciousness trick. Yet for most people, human conscious experience is not a trick; it is a very real thing, something that is managed by a part or system in the brain, and the hunt is on to find it. Since humans possess advanced cognitive processing that allows for the development and utilization of new technologies and for the making of inferences about the beliefs and desires of others, do human brains possess something that animals do not?

A recent comparative study looked at the neuropil volume in different areas of humans' and chimps' brains.[43] Neuropil comprises the brain areas that are made of connections: a mixture of axons, dendrites, synapses, and more. The prefrontal cortex—the brain area in humans involved in decision making, problem solving, mental state attribution, and temporal planning—has a greater percentage of neuropil than is found in chimp brains, and the dendrites in this region have more spines with which they connect to other neurons than do other parts of the brain. This anatomical finding suggests that the connectivity patterns of the prefrontal neurons may contribute to what is different about our brains. Interestingly, corvids have a relatively larger forebrain than most other birds, especially the areas that are thought to be analogous to the prefrontal cortex of mammals.[44] Yet, as we shall see, while this way of thinking may explain increased abilities, it is not going to get us to the goal of understanding how consciousness is enabled. Backsliding into the assumption that there is special sauce or a special brain region that gives us conscious experience is a nonstarter.

Where Is Consciousness?

We have to shift gears. We have to rid ourselves of the notion of the special sauce, the special place and thing. We have to think about the aggregate of largely independent modules and how their organization gives rise to our ever-present sense of conscious experience. As cognitive scientists, we get too fixated on the idea that consciousness is a phenomenon separate from our other psychological processes. Rather, we should be thinking about consciousness as an intrinsic aspect of many of our cognitive functions. If we lose a particular function, we lose the consciousness that accompanies it, but we don't lose consciousness altogether.

An early clue that consciousness is not tied to a specific neural network came from my own studies on split-brain patients. While there are more neural connections within a half brain than between the two halves, there are still massive connections across the hemispheres. Even so, cutting those connections does little to one's sense of conscious experience. That is to say, the left hemisphere keeps on talking and thinking as if nothing had happened even though it no longer has access to half of the human cortex. More important, disconnecting the two half brains instantly creates a second, also independent conscious system. The right brain now purrs along carefree from the left, with its own capacities, desires, goals, insights, and feelings. One network, split into two, becomes two conscious systems. How could one possibly think that consciousness arises from a particular specific network? We need a new idea to cope with this fact.

Consider, too, what the conscious experience is like for the split-brain patient who wakes up from surgery, and each hemisphere now doesn't know about the other hemisphere's visual field. The left brain doesn't see the left side of space, and the right brain doesn't see the right side. Yet the patient's speaking left hemisphere does not complain of any vision loss. In fact, the patient will tell you he doesn't notice any difference in anything after the surgery. How can this be when half the visual field is gone? Like a patient with spatial hemi-neglect, the speaking left hemisphere never complains that it has lost half its visual field. The modules that are responsible for reporting the loss are over in the right hemisphere and can no longer communicate with the left. The left hemisphere neither misses them nor is aware that they were ever there. The memories of having had that visual field are also gone from the left hemisphere. The whole conscious experience of the left visual field is now enjoyed only by the right hemisphere and completely disappears from the left hemisphere's experience. What does this tell us about consciousness?

After being weaned away from the idea of a single "conscious" module, we can begin to narrow in on what consciousness actually is. We know that local brain lesions can produce various specific cognitive disabilities. Yet such patients are still aware of the world around them. The patient with a severe spatial neglect is not aware of the left half of space, but is still aware of the right.

What if conscious experience is managed by each module? Lose a module to injury or stroke, and the consciousness that accompanies that module is gone, too. Remember: patients with hemi-neglect aren't conscious of one-half of space because the module that processes that information is no longer working. Or, if the modules responsible for locating oneself in space are not being integrated properly, conscious experience is deeply affected, and one ends up with the feeling that someone else is there just over your shoulder. Or, take people with Urbach-Wiethe disease, which leads to deterioration of the amygdalae: they no longer experience the emotion of fear. One such patient, although she has been extensively studied for more than twenty years, has no insight into her deficit and frequently finds herself in scary situations.[45] It seems that because she doesn't have the conscious experience of fear, she doesn't avoid those situations.

This idea that consciousness is a property of individual modules, not a single network a species might have, could explain the different types of consciousness that exist across species. Animals are not unconscious zombies, but what each is conscious of differs depending on the modules it has and how those modules are connected. Humans have a rich conscious experience because of the many kinds of modules we possess. Indeed, humans might well possess highly developed integrative modules, which allow us to combine information from various modules into abstract thoughts. It is difficult to decipher how consciousness arises in humans, but thinking about consciousness as an aspect of multiple functioning modules may guide us to the answer.

Even so, if consciousness is an aspect of several different cognitive domains, how is it that people with an intact corpus callosum still experience the world as a single entity rather than a world of randomly presented snippets at any given moment in time? To understand this, we can relate the brain's processing to a competition. Modules vary in the amount of electrical activity they possess from moment to moment, with the result that their contributions to our conscious experiences vary. The idea here is that the most "active" module wins the consciousness competition, and its processing becomes the life experience, the "state" of the individual at a particular moment in time. Imagine that you are on a beach watching an exotic bird fly through the air. At that moment, the visual thrill, the sight of the bird and its colorful feathers, has won the conscious-experience rivalry. The next moment the competition has been won by the call of another bird, and the next by a surge of curiosity, so you turn your head to locate the source of the sound. All of a sudden a sharp pain in your foot has priority, which immediately causes you to look down to see a crab clamped onto your toe. At every moment in time, your single conscious experience is the cognitive aspect that is most salient in your external or internal environment; it is the "squeaky wheel." All the various competitive processing was performed by different modules. How does this all work?

I propose that what we call "consciousness" is a feeling forming a backdrop to, or attached to, a current mental event or instinct. It is best grasped by considering a common engineering architecture called layering, which allows complex systems to function efficiently and in an integrated fashion, from atoms to molecules, to cells, to circuits, to cognitive and perceptual capacities. If the brain indeed consists of different layers (in the engineering sense), then information from a micro level may be integrated at higher and higher layers until each modular unit itself produces consciousness. A layer architecture allows for new levels of functioning to arise from lower-level functioning parts that could not create the "higher level" experience alone. It is time to learn more about layering and the wonders it brings to understanding brain architecture. We are on the road to realizing that consciousness is not a "thing." It is the result of a process embedded in an architecture, just as a democracy is not a thing but the result of a process.

5.

THE BEGINNINGS OF UNDERSTANDING BRAIN ARCHITECTURE

There are a great many things about architecture that are hidden from the untrained eye.

—Frank Gehry

MAGINE YOU ARE the budding young scientist that your parents think you are. They give you an old alarm clock for Christmas and say, "Okay, Mister Smarty-Pants, take it apart and put it back together, and while you are doing that, tell us how it works." That would be easy enough. There are only a small number of parts—wheels, gears, springs—moving together in an architecture to produce a function, and we know what the function is. It would be much more difficult if you had no idea what the function was and only had the parts in hand.

For those of us studying the human brain, the problem rests in determining how 89 billion neurons connect to one another to allow us humans to strut our cognitive stuff. Brains are dissected, stained, poked, mapped, and eavesdropped upon. Vast amounts of data have been carefully collected, injured patients have been carefully studied, and mental feats of exceptional people have been examined for the underlying, mystifying magic that we are seeking to understand. Twenty-six thousand brain scientists gather once a year at the Society for Neuroscience meeting to exchange their data and their thoughts, and still the field is searching for a framework to place all of this information in. Why is it so elusive? What are they missing? There must be another dimension of the problem that needs to be captured. In the mid-twentieth century, the theoretical biologist Robert Rosen suggested to his daughter one possible dilemma: "The human body completely changes the matter it is made of roughly every eight weeks, through metabolism, replication and repair. Yet, you're still you—with all your memories, your personality. . . . If science insists on chasing particles, they will follow them right through the organism and miss the organism entirely."[1]

Rosen's comments hint that organization must be independent of the material particles that make up a living system. Indeed, the structural components and the function of a brain are only part of the story. A third, and often overlooked, component is necessary to relate the structure of a system to its function. Missing is how the parts are organized, the effects of any interactions between the parts, and the relations with time and environment. This was dubbed *relational biology* by Rosen's professor Nicolas Rashevsky, a theoretical physicist and mathematician at the University of Chicago. These ideas have filtered down to researchers in electrical engineering and systems biology, but are generally unknown or ignored by molecular biologists and neuroscientists, even fifty years after Rosen's warning.

My own introduction to this alternate way of thinking about brain organization came from John Doyle, a professor of control and dynamical systems, electrical engineering, and bioengineering at Caltech. First lesson from Dr. Doyle: learning about the parts can only take you so far. A school has places for reading and eating, for washing your hands and storing stuff. So, too, does a house. Yet a school is not the same as a house; they serve different functions, and the flow of people through them is vastly different. A major difference is in the organization of the parts, its architecture. The Hungarian-British polymath Michael Polanyi explained that a "machine as a whole works under the control of two distinct principles. The higher one is the principle of the machine's design, and this harnesses the lower one, which consists in the physical-chemical processes on which the machine relies."[2] The machine's design restricts nature in some way in order to harness it to do a particular job. For example, your coffee maker is made up of aptly designed parts that fit together such that it produces a cup of coffee. Polanyi calls these restrictions the imposing of *boundary conditions* on the laws of physics and chemistry. He shows that organisms are systems that share this characteristic with machines: "The organism is shown to be, like a machine, a system which works according to two different principles: its structure serves as a boundary condition harnessing the physical-chemical processes by which its organs perform their functions. Thus, this system may be called a system under dual control."[3] The design that Polanyi is referring to is the organism's architecture, and that is key to understanding the mind/brain complex. This insight is crucial.

The Architecture of the Complex

Doyle is a master at thinking through how complex systems—things like a Boeing 777 or your brain, both of which are composed of many interacting parts—can work efficiently, quickly, and safely rather than exploding, crashing, or screeching to a stop. It should come as no surprise that the word "complexity," like "consciousness," does not enjoy a universally accepted definition. For present purposes, we can focus on three dimensions of complexity in a system. A system is complex if it has a large or diverse number of (a) components, (b) interconnections and interactions, and (c) resultant behaviors, some predictable, others not so much. Engineered systems have begun to have almost biological levels of complexity.[4] For instance, the Boeing 777 has, by Doyle's estimate, 150,000 different subsystem modules, organized into complex control systems and networks, including roughly 1,000 computers that fly the plane. While the components of advanced technological systems and highly evolved biological systems are obviously different, they share similarities in their organizational architectures.[5]

Ordinarily, the term "architecture" brings to mind the art and science of designing buildings and other structures like bridges and freeways, their style of design (Baroque, Art Nouveau), and their method of construction (rammed earth, glass and steel). Perhaps thoughts of Brunelleschi or Palladio come to mind. But architecture also means the complex structure of something. That something is not necessarily a building and may or may not be physical. It could be the command structure of a government, or the pathways of the Internet, or the neuronal networks in your brain. Architecture at its fundamental level is about *design within the bounds of constraints*. These are Michael Polanyi's boundary conditions: boundaries imposed by a comprehensive restrictive power.[6] For a building this means working within the limits of the materials used (grass, mud, wood, brick, stone, steel), the site to be built upon (regions notorious for fires, flooding, earthquake, or hurricanes; flat or steep; tropics or tundra), the function of the building (home or opera house or gas station), and, of course, the desires of the owners (Aristotle's final cause), and so forth. For your brain and nervous system, architectural constraints include energy costs, size, and processing speeds.

Complex biological and technological systems share a highly organized architecture: that is, the components of the system are specifically arranged in a manner that enables their functionality and/or robustness. For a simple example, the cotton fibers in a garment have a highly organized architecture, which makes the cloth functional for clothing.[7] It is robust to the wear and tear that a garment sustains. In contrast, randomly mashing together the same cotton fibers results in paper, which is not robust to that same wear and tear. The similarity of architecture in organized, complex systems suggests that they all share universal requirements. They are designed to be "efficient, adaptive, evolvable, and robust."[8]

The Robust, the Complex, and the Fragile

Understanding that animals, both large and small, are not a whole lot different in design from your BMW or pickup truck allows for clearer thinking on the nature of how biological tissue does its thing. Doyle and his colleague David Alderson argue that complexity in highly organized systems is not accidental. It arises from design strategies (whether created or evolved) that create robustness or, as Darwin would have called it, fitness.

Doyle and Alderson define robustness in this way: "A [*property*] of a [*system*] is *robust* if it is [*invariant*] with respect to a [*set of perturbations*],"[9] with brackets indicating that each of these terms has to be specified. Take an casily understood complex system to illustrate robustness: clothing. Suppose you are packing for a trip to see the northern lights. It will be in the far north in the winter, and you want to stay warm. Down [property] clothing [system] for your cold-weather trip may be a robust choice, since it keeps you warm as the temperature drops. If there is a drenching rain [a perturbation that was not specified], however, and your jacket gets soaked, the down will not keep you warm. Though down is (relatively) invariant to cold temperatures, it is not invariant to wetness: robust to some conditions but fragile to others. Or if you choose "makes me look svelte" as your [property], and [weight gain] as your perturbation, then you will end up sporting skimpier choices and looking slim in a blizzard (robust to weight gain), but your clothing will not be invariant to the perturbation of drops in temperature.[10]

Each feature that adds robustness to a system protects that system from some internal or external challenge. Each step toward robustness also increases complexity. Sadly, no added feature is robust for all contingencies. Each feature will add its own Achilles' heel into the system, a new vulnerability to a new unforeseen challenge. So, once that is discovered, another feature must be added that combats that new fragility. However, with the next feature comes yet another fragility that needs to be safeguarded against. Each safeguard adds complexity, which demands further complexity.

Your property (or feature) trade-offs inevitably render your system's behavior robust to some perturbations but fragile to others. "Robust yet fragile" features are a characteristic of highly evolved complex systems. The concept of robustness, always accompanied by fragility, is found all over the place. One of my favorite biological examples comes out of studies on brain development.

It is obvious that neural connections are important for normal brain function. Neurons in one place have to have connections to another place in the brain for the ultimate coordination of activity to produce behavior. Evolution seems to have robustly assured this by making sure there is a huge overproduction of neurons during development. Instead of structure A sending just the right amount of neurons to structure B, it sends way too many, to ensure robustness. Mother Nature has figured out a way to get rid of the extra neurons through a process called "pruning." With appropriate input from the environment, the unnecessary neurons die off, and an acceptable number connect the two structures at the end of the developmental period. But, of course, here comes the fragility. Frequently the pruning is overdone. Indeed, evidence has accumulated that developmental mistakes in pruning lead to autism[11] and schizophrenia.[12] "Robust yet fragile" is everywhere, and the concept is central to realizing how the brain is organized.

The Universal Design Strategy

For Doyle, it is clear that most biological systems have a "layered architecture." Any attempt to understand conscious experience, therefore, must build on a firm knowledge of how the brain is organized into layers. Many researchers, who have been conditioned to using cognitive models, might not initially see the difference between "levels" and "layers." With levels, processes are sequential (or, as electrical engineers would say, "in series"), while with a layered architecture, processing goes on simultaneously ("in parallel"). When processing through levels, all the steps are performed one after another, like a baton relay. You need one level to finish before the next one up can start. Processing in layers, on the other hand, can have all the runners leave at the same time and go different places. This change in architecture makes for big differences.

Layering architecture is the key design strategy to ensure robustness and functionality for both organized technological and biological systems. It's simple, necessary, powerful, and hugely beneficial. For example, good design in both technical systems, such as a Boeing 777, and biological systems, such as our brains, is done in such a way that the users are largely unaware of all of the hidden complexities.[13] We just get on the plane, tilt the seat back, and pull out a book or order a drink. We don't think about the plane's 150,000 subsystem modules and what they are doing. And neither do the pilots. We don't even know there are 150,000 subsystem modules. And if you skipped the last chapter, you may not even know what a module is. Similarly, most of us don't give our brains much thought unless something malfunctions. The complexity of our layered brains is so well hidden that after twenty-five hundred years we are still trying to figure it out. In both the 777 and our brains, the system's architecture hides the complexity. So what is layered architecture, anyway?

An engineer's goal is to design and build stuff that works efficiently, effectively, and reliably. Hard enough when you are building a pergola* over your patio—now imagine the Sydney Opera House. On such a project, not only do the parts of the building have to come together to work efficiently, effectively, and reliably, but so do the engineers who have to work together to produce it. One person doesn't design every aspect. Yet if the wrong design strategy is used for organizing the engineers, this "many cooks" approach can be a recipe for disaster.

In fact, the engineers who design complex systems are themselves a complex system and are organized in a similar manner. Let us consider different strategies for designing a 777 and its operation. One strategy is that,

---

*One of my friends who read this remarked, "What's a 'pergola'? This is not a term us Lower East Side NYC guys have ever heard." I am going to let you look it up on Wikipedia, just like my friend ended up doing. That way you can see a picture, too.

in order to design an assigned chunk of the plane, every engineer must understand every other engineer's work. Then, once the different chunks are designed, every component must depend on all the other components to function properly; that is, everything must be integrated serially. This would mean that the engineer designing the seats would need to know all about the engines, lift and drag, window glass, pressurizing systems, and so forth, and would have to fully integrate the seat's function with them. Not only would the construction of the plane take longer, be more expensive, and require more multidisciplinary experts, but it would open itself up to more errors: a seat that didn't tilt back wouldn't just annoy you—it might cause the plane to nosedive.

The better strategy is to independently design independently functioning components (layers or modules). The designers work with only the information they "need to know." All the rest of the information is kept from them. In the engineering world, this is known as *abstraction*, the removal of unnecessary detail (more abstraction = less detail). "Layers of abstraction" refers to what information is available and what is hidden. Layers of abstraction do not always demand hierarchy or even fundamentally different components. A world atlas has multiple layers of abstraction, yet each layer is in the same form. Page one has a map of the world. You'll see the oceans, continents, perhaps major rivers and mountain chains labeled. However, most of the information has been left out: no countries, cities, roads, creeks, or hills. Flip the page and you have the next layer of abstraction, a continent with various countries, capitals, rivers, and mountains. Flip it again and you will find a map of a single country in more detail, with major roads and smaller cities highlighted. At each level of abstraction you are seeing more and more detail; less information is being hidden. But more information isn't always better: if you are only interested in the relative size of the different oceans, you don't need to know that there is a footpath between Roussillon and Fontaine de Vaucluse.

In a complex system, however, information isn't just hidden. Each layer has a completely different form. To transfer between layers, the necessary information must be virtualized, that is, abstracted to a specific layer. So in the 777, the seat engineer is given only the information needed to make seats: a set of standard measurements that allow for flexibility in the seat design but constrain the process such that all seats will fit together across all 777s. No information is given on aerodynamics, none on fuel, not even the number of seats the plane will have. I have personally noticed that, apparently, airplane seat designers have not been given the information that there are people who are over six feet tall.

Here, seats become interchangeable modules. The seat's functioning doesn't affect the plane's flying ability. So the seat engineer knows more about the plane than you do, but less than the engineers who design the body of the plane. Meanwhile the engineers responsible for the plane's turbines needn't know anything about the seat design but have a host of other information instead.

Mother Nature figured this out a long time ago and uses the same strategy in evolved organisms. The various systems in your brain have evolved to run independently. For example, your auditory system runs independently from the olfactory system. It gets no information about odors and doesn't need any to process sound information. You can lose your sense of smell and still hear bees buzzing.

In a layered architecture, each layer in a system operates independently because each layer has its own specific *protocols*, the set of rules or specifications that stipulate the allowed interfaces, or interactions, both within and between the layers. Consider the seat engineer again. The engineer can go hog-wild on seat design as long as it stays within the set of standard measurements, the protocol for the seat layer. A protocol for a layer constrains, but it allows flexibility within those constraints.

Each layer in a "stack" of layers processes the output it receives from the layer below, according to its specific protocol, and passes the result on to the layer above and/or back to the layer below. The layer above does the same, according to its own specific protocol, which may be similar or completely different, and passes its processing results on up to the higher layers. No layer "knows" what the previous layer received as input or what processing occurred. It doesn't need to, so that info is hidden (abstracted). The protocols allow each layer to interpret only the information that it receives from its neighbor layers. Information produced from the processing within the layer can then be sent up or down. Here's a catch: once a layered architecture is created, information cannot skip layers. Thus, a sixth layer would not be able to interpret the fourth layer's output because it has no protocol to decipher it, hence the need for a mediating fifth layer. The purpose of each layer is to serve the layer above it while concealing the processes of the lower layer.[14]

Here's a simple example of layers with protocols: Imagine you are at a party with many international guests. You want to speak to the Chinese woman who apparently knows your sister. You only speak English, but your partner speaks English and also French. The Chinese woman only speaks Mandarin Chinese; however, her husband speaks Chinese and French. Each person becomes a layer in the translation stack, with each layer using his or her own protocol to transform the input information into its output form. You have the protocol for English and output English to your partner. Your partner takes your output and uses the English and French protocols to output French to the next layer, the husband. He shares the protocol for French but also has the protocol for Mandarin, which he outputs to his wife. She can also send information back down the layers to you, but neither you nor your Chinese guest can skip the intervening French layers. Information can flow up and down the layers, but it must be processed by layers with the appropriate protocols and handed off to the next layer. Creating an English-to-Mandarin layer of your own won't happen over the course of one dinner party.

However, you can create an English-to-Mandarin protocol if you so choose. Plenty of people have. This is called learning a language, and you can do it, as can machines. In fact, computer scientists have been using layered architectures for years, especially in the world of artificial intelligence. Rodney Brooks, the ingenious computer science professor at MIT, had an idea of "subsumption architecture" that dominated the field of robotics for many years.

On this topic the dictionary is not helpful,* but the idea is simple enough. A system, such as a person or a computer or a robot or a library, has some knowledge stored in it. Along comes some new information adding to the overall knowledge of the system in question. Ideally, that information is "subsumed," or absorbed by the existing thing without disturbing matters. This architecture is exactly what a robot needs.

Brooks knew well that robots failed the subsuming-new-information test because, twenty years ago when he presented this idea, robots would seize up if they ran into a marshmallow, or really any object that had not been previously specified in the robot's programming to avoid. They couldn't adapt to a changing environment. If the robot had a subsumption architecture, however, it would allow for incremental changes by adding new layers to the existing layers, one at a time. Each could inject its influence into lower layers and would thereby be subsumed by the larger architecture. As Harold Pashler, in the *Encyclopedia of the Mind*, summarized: "A key idea is that the system as a whole does not construct integrated representations of the world; rather, sensory signals are processed differently at each level to implement relatively direct and behavior-specific mappings between sense data and the motor signals required to control the robot's actuators."[15] That means there are very specific systems all over the place precisely managing day-to-day challenges to the robot. They are fast and efficient and useful. There is no central system changing the response pattern of the robot to meet every new challenge. Instead, a specific instruction is added that deals with that challenge, and, over the course of time, engineers have added more and more instructions to deal with more and more perturbations in the environment. There is not one big thing trying to figure everything out. There are new

---

The American Heritage College Dictionary: Subsumption: "The act of subsuming"; "Something subsumed"; or "The minor premise of a syllogism," where a syllogism is "A form of deductive reasoning consisting of a major premise, a minor premise, and a conclusion."

layers added as things come up. It sounds like all those modules fit together into a layered architecture. In fact, a module can be a layer itself or a layer may be made up of a group of modules. When we spoke of integrative modules at the end of chapter 4, these were processing modules making up a layer high in a stack. That layer received information from a previous layer and processed it, according to its protocol, to produce something more complex, perhaps even theory of mind or self-awareness!

Layering allows for flexibility. Implementing updates to a layered system is simple because changes need to occur only at a specific layer without the need of altering the other layers. And, when something does go wrong, the source of the error is identifiable. The entire system doesn't have to be repaired or junked, just the layer or parts of a layer that aren't working. With your layered clothing, if your shirt rips, you can swap it out—you don't have to change your pants in the process. In the case of your brain, you may not be able to swap out parts so easily, but you may not lose the whole system if something goes wrong.

The Neuroscience of Layering

The beauty of layered architecture is that it solves a problem for the user of complex systems by hiding information. On your iPhone, that top layer is known as the application layer; it makes it so we don't have to know or understand how the other layers of the system work. It's not like you'd want to have to figure out your iPhone's memory allocation protocols anytime you want to send a group text or snap a photo. Likewise, we should count our lucky stars that we don't have to understand how the brain works to use it. We don't know how our lunch is turned into energy—we just eat it and go. And the same is true in our mental life. We don't have the slightest idea how we do anything. Point to your nose. Do you know how that all happened? Knowing how to do it in terms of generating the neural messages to your muscles is completely outside our realm of awareness and knowledge. Just as the pilot in a 777 works a computer program to fly the plane, we work our minds, the brain's application layer, to produce actions, which is to say, our behavior. But do real brains actually have a layered architecture? Or is this a formalism that does not truly find a home in real biological systems?

As is frequently the case, the discovery of a new perspective or idea usually results in the discovery that other scientists have had similar ideas, sometimes years before. How many times must we learn that human ideas are exactly that—human ideas, which lots of humans have had throughout history? In our case, we look to Tony Prescott, Peter Redgrave, and Kevin Gurney at the University of Sheffield. All three are well-versed in neuroscience, robotics, and computer science, which they use with unending cleverness. In their seminal paper on layering,[16] written almost twenty years ago, they directed us to where we are today. The journey starts with John Hughlings Jackson, the great nineteenth-century British neurologist. Jackson was by all accounts a brilliant physician—what we would call a first-rate jockey. Unfortunately, he rode a third-rate horse when it came to taking up the pen: his writings are almost impenetrable. Luckily, a few jockeys riding first-rate horses have made his work understandable to the world.

Darwin had inspired the scientific and medical world, and Jackson was all in. The brain, through natural selection, was a sensorimotor machine, and each species has its own set of evolved abilities. In the human, the higher layers are the most sophisticated at coordinating an act, but both the higher layers and the lower layers have the basic abilities wired into them. For example, a cat or a rat with its cortex removed can still show various motivated behaviors, such as walking, grooming, eating, and drinking. But without the higher layers, certain more complex behaviors go missing. As Prescott and his Sheffield colleagues put it:

> [Jackson] divided the nervous system into lower, middle, and higher centers, and proposed that this sequence represented a progression from the "most organized" (most fixed) to the "least organized" (most modifiable), from the "most automatic" to the "least automatic," and from the most "perfectly reflex" to the least

"perfectly reflex." This progression sees an increase in competence in a manner that we might now understand as a behavioral decomposition—higher centers are concerned with [the] same sort of sensorimotor coordinations as those below, though in a more indirect fashion.[17]

Jackson immediately saw the implications of his layered view and suggested that there should be *dissociations*, a term he introduced to neurology, such that specific brain lesions should produce particular kinds of behavioral deficits. Knock out the top layers, and only the bottom layers can respond. And respond only with their limited level of capability, just as described for the decorticated cat.

Evolvable Layered Systems

Of course, all of this pathbreaking work forced the question of whether or not the brain develops in a layered way over an evolutionary timescale. Are brain areas added slowly but surely, and is there such evidence coming out of studies on comparative anatomy? There is indeed, and this is where Prescott shines. He lays out a long, complex, and fascinating story: the evolutionary process for modern nervous systems in all vertebrates started more than 400 million years ago with the basic plan of the spinal cord, hindbrain, midbrain, and forebrain. As the millennia rolled by, the forebrain added modules and layers that brought new functionality, not simply elaborating the old. For example, as the limbs became more capable of manipulating things, new neural acreage was needed for modular layers to supply control of those new peripheral manipulators, otherwise known as fingers. These new neuronal pathways are clearly present in vertebrates with digits, but are not present at all in vertebrates without them. And just as Jackson predicted, lesions to the forebrain disrupt one set of modules reigning over fine motor control of the hand, but don't disturb others concerned with more basic motor control of the arm.

The capacity to evolve is a good thing for any given population of animals to have, as it serves as the basis for adapting to new challenges. It is defined as an organism's capacity to generate phenotypic variations (observable traits) that can be inherited.[18] If a trait is selected by natural selection, it will be passed on to the next generation. A well-known example from the Galápagos Islands is the range of beak sizes, from small to large, found in the ground finch.[19] Yet one of the puzzles posed by Darwin's theory of selection of heritable variation was: Just exactly where was all that variation coming from, and how was it generated? The stock reason offered—that it was mostly due to random gene mutations—can take up some of the slack, but not all of it. It has been a puzzle for biologists for years.

Along come the Harvard biologist Marc Kirschner and his Berkeley colleague John Gerhart.[20] They wondered whether modern creatures have cellular and developmental mechanisms with the characteristic of what is called *evolvability*. That is, do they have the ability to generate heritable phenotypic variation? And is the characteristic of evolvability itself under selection pressure? That is, will biological systems that produce more phenotypic variations that can be passed on to their offspring be more likely to win evolution's arms race?

Throughout the animal world, there is great diversity in body shape, tissue organization, development, and physiology. At the same time, many core processes, such as biochemical and cell signaling pathways, as well as the circuits that regulate the expression of genes, are the same throughout. We animals share some core processes with plants, fungi, and slime molds: for example, we use the same enzymes to regulate cell division. We share other core processes—metabolism and replication—with life forms down to bacteria. Why? Because we share many of the same genomic sequences. While some biologists think that these core processes constrain evolution, Kirschner and Gerhart do not. In fact, they think quite the opposite. They think that the reason why we share so many core processes, why they have stuck around for the past 530 million years, and why they have proved successful is that they not only bestowed flexibility, rather than limited it, but also allowed successful variations to be passed on to offspring. The flexibility these processes bestowed was phenotypic variation in the processes that were fragile to environmental changes. Thus, the core processes were a constraint that provided evolutionary flexibility in the face of environmental uncertainty.

So, if you think that all sounds like protocols in a layered architecture, "constraints that deconstrain," you are on the money. Doyle and Alderson lament that this role that layered architecture plays in producing variation has not been fully appreciated by most biologists. Perhaps layered architecture, which is ubiquitous in biological systems, evolved because its ability to generate variation within a set of constraints was robust against the competition and was selected for: layer or die!

The Binds That Make It Free

Recognizing a layered architecture is one thing, but figuring out how the layers talk to one another is another. A wide diversity of information coming into a layer must be processed and converted to a form that is interpretable for the next layer. The major constraints of a layered structure occur in the processes that connect the layers.[21] This feature of the architecture is cleverly visualized as a bow tie or an hourglass, with the protocol being the constraining knot, and the inputs and outputs fanning out. In our previous example of the seat engineer, the protocol for the seat layer, those measurements were the knot in the bow tie. What was coming into the protocol were all sorts of possible building materials, shapes, colors, and so forth. The outputs could be a wide variety of seats of various materials, designs, and colors, but they all had to conform to the protocol measurements and function. Overall, the system is constrained, but also *deconstrained*, a made-up word.[22] A set number of inputs can become several different kinds of outputs, because now there are multiple ways to complete a layer's task. When you think about it, it is almost magic. Looking at a suitably layered robot, it almost seems that this tightly wired-up system with only set responses is actually thinking in the warm and fuzzy sense. The architecture has allowed the system to be more flexible.[23]

The fact that a layer's protocol both constrains and deconstrains a system is crucial. I want to pound this in, so here is another example. Think back to your layered outfit and the whole range of possibilities that are available for each layer. For instance, the warmth layer (with its constraining protocol: must trap body heat) has all types of garments as inputs and many possible output outfits. It could be a bearskin cape and wool pants, a merino wool coat with a cashmere sweater and sheepskin pants, or a polypropylene fleece jacket with a mink vest and rubber wet-suit pants. It can be zippered or a pullover, overalls or a separate top and bottom. It can have a turtleneck collar and elasticated wrists and ankles, or not. It can be any color or size. Though the warmth layer has a protocol that constrains it (must trap body heat), the protocol also deconstrains it in any number of ways, which allows a huge variety of choices. In Darwinian terms, we have selection from variation. We can see that this flexibility allowed the layer to evolve from an animal-skin cape to a polypropylene fleece, zip-fronted, size L magenta jacket with a hood. And pockets. This illustrates what could be the most important feature of layered architecture: it enables change over large timescales, from the outfits of Fred and Wilma Flintstone to an Armani suit and a Valentino gown.

Despite the flexibility of layered architecture, there are drawbacks. Let us take the above permutations of warm clothes and now demand that the output be "fashionable." This is a more difficult problem. The more specific the protocol, the greater the constraint. In this respect, a unified system that does not operate in layers tends to work more efficiently, because it does not have separate protocols for each of its functions. Again, a protocol is a set of rules, or specifications, that stipulate the allowed interfaces, or interactions, both within and between the layers. It would be easier to just design one jumpsuit of some expensive material that keeps you warm, dry, and comfortable all in one garment that is flexible and lightweight. It could squish into your carry-on bag and would be quick to put on. It even might make you look svelte and be fashionable! Then for the rest of your life you'd just need this one garment. Brilliant!

A unified functioning structure, however, is not ideal for a complex system because one minor malfunction would crash the whole kit and caboodle, and updates could not be easily applied. You rip the leg of the ideal jumpsuit on a nail, and the whole suit unravels. At least with the damaged layered system you still had some backups, and it is easier and cheaper to repair or replace. Or what if a better material for breathability comes along? You would have to junk the entire garment and get a new one to take advantage of it: costly. It takes more time, energy, and resources to maintain a unified system; that is, even though it may be more efficient, the trade-off is that it is more costly and not as robust. Because each layer can provide a wide range of diverse functions, the system has greater flexibility as a whole, giving it a great advantage when facing a changing environment. This type of layout is ideal in an evolutionary sense because the number of vulnerabilities in the system is limited, while the opportunities for diversification are abundant. As an environment changes over time, such systems can adapt more readily. Overall, a layered architecture is ideal for complex systems because it is easily fixable, less costly, more flexible, and evolvable.

However, layered complex systems are not immune to protocol malfunctions. When the system breaks, is under siege, or even is hijacked, failures—some catastrophic—can result. For instance, if the seams of your wool pants unravel due to a protocol failure in the sewing layer, the pants are no longer functional: you might as well be wearing a hula skirt. If, in the biological system that is your body, the protocols of your immune system are hijacked, you can end up with an autoimmune disease. Because complex systems have many components and layers of subsystems, interactions may result that are very difficult to predict. For example, a small fault in a component that doesn't greatly affect the component's local performance may be magnified when it interacts with other components and affect the overall system performance. This unpredicted interaction may lead to a system failure.[24]

Damage to the protocols can compromise even the most robust system, but the advantage is that, on the whole, there are few major attack points. Even better, the system may be able to limp along with its deficit, whereas in a unified system an attack on any component can compromise the entire system.

Why can't we just work these bugs out of the system? Why can't we get rid of these attack points? The problem is that new strategies that solve one problem invariably introduce new vulnerabilities, which then have to be patched up. Increasing robustness is an arms race resulting in systems adding more and more layers, becoming more and more complex. The evolution of our complex brain and body from that first chemical stew, and of the 777 from nuts, bolts, and Orville Wright's bicycle, has simply been the result of an arms race of adding layer after layer of robust features to combat fragilities. It is like the Red Queen in *Through the Looking-Glass*, who runs faster and faster to stay in the same place. So are we doomed? One counterstrategy is to introduce redundancies into the system.

There Are Multiple Ways to Skin a Brain

While the protocol of a layer constrains its output by limiting the possibilities, it doesn't dictate which of the possibilities will be its output. Constraints that deconstrain are not the same as causality. A constraint may limit the number of outcomes, but it does not cause the outcome. If you are dressing for a party, your outfit will be constrained by the clothes in your closet (and possibly what you consider to be socially acceptable party wear), but that constraint does not dictate that you wear a specific outfit. There are still plenty to choose from. Protocol constraints that deconstrain do not dictate the outcome. Making the assumption that they do can cause a problem for those who are unaware that they are dealing with protocols within a layered architecture. They may think they can look at a behavior and predict the neural firing pattern, that is, the "brain state," that produced it. The error in this way of thinking is well demonstrated by the work of the brilliant neuroscientist Eve Marder,[25] who studies lobster guts.

Marder was studying the "digestive layer" of the lobster, examining the contractions of the gut. She isolated and studied every single one of the neurons and synapses, down to their neurotransmitter effects, that are involved in the lobster's gut motility. Just as you have 1 billion possible outfits in your closet (by combining your clothes in all sorts of weird ways, such as socks on your hands and skirts over Wranglers), she found that there are 2 million possible network combinations in this tiny gut. But, just as with your outfits, protocol constraints cut down on the possible outputs: only a small percentage of them work. You don't wear your boxers outside your slacks or your jacket under your cocktail dress, though you could. Just as with your outfits, a small percent of a billion or a small percent of 2 million is still a lot of variation. In fact, 1 to 2 percent worked: 100,000 to 200,000 tunings of that handful of neurons will result in the exact same behavior at any given moment. There were multiple ways to complete the task of the motility layer, just as there are multiple ways to complete the task of putting an outfit together. This is an example of *multiple realizablity*, the notion that when it comes to neurons firing, there is more than one way to skin a cat. That is, the same mental property, state, or event can be implemented by different neural firing patterns. This may seem like a waste of evolutionary time or biochemical energy. However, what this means is that if one pathway breaks down, another can pick up the slack.

Layered systems minimize resource costs by developing adaptable components that can serve multiple purposes. For example, in the biochemical layer of the brain there are many proteins that serve multiple functions involved with signaling pathways and feedback loops to regulate the system across various layers.[26] The system saves energy, since it does not need to develop several unique components for each layer.

Another cost cutter is the damage control provided by parallel processing. For instance, in multicellular organisms there is the cell layer, in which cells are individually metabolizing, humming away independently, and there is a tissue layer made up of those cells working together, performing the task of the tissue following the tissue protocol. Each cell in the cell layer carries out its own protocols, which may be identical to those of other cells within the tissue, but each is humming away independently. If a cell is damaged, then resources are needed only to revive a single cell rather than the entire tissue. Additionally, if a cell is destroyed beyond repair, then the tissue will most likely still function, since the absence of a single independently functioning cell is usually negligible. As in any layered system, however, the system can become jeopardized if a core component that transfers information from one layer to another malfunctions. In the case of biological tissue, if the kinds of proteins that connect cells together are defective, then information may not be able to transfer from the individual-cell layer to the tissue layer, and the entire system could shut down. While not perfect, the layered design of biological systems is advantageous because it minimizes the number of points in the system that, if attacked, would result in catastrophic damage, and it limits the effects of attacks at other points.

Kirschner and Gerhart propose that traits with the most flexibility paved the way for organisms with more complex development, since more dynamic systems tend to be less vulnerable to lethal mutations. Populations of organisms are able to evolve because of the phenotypic variation that arises from a layered architecture. Thus, the "evolvability" of layered systems may have greatly contributed to successful survival over time.

From Chromosomes to Consciousness: A Layered Architecture

We are starting to walk on thinner ice here. While it is understood that complex biological systems have a layered architecture, the overall roles of the layers, all their functions and their dynamics, are not yet known. Some systems, such as bacteria, are better understood than others, such as—you guessed it—the brain. In a biological system, what Doyle calls the *composition layers*, the evolutionarily oldest layers, are the most basic nuts and bolts of the system, built up from subatomic particles, to atoms, to molecules, each with their own protocols. Other layers would include one with protocols describing interactions within and between the molecules; another with a protocol that can produce dynamic interactions, that is, those that produce change or progression; and control layers with protocols that use feedback to adjust a system's response. Control protocols guide the system's behavior in the face of various internal and external perturbations. And one thing we do know is that the control systems in the maturing human brain take longer than many other systems before they are fully up and running. If you aren't convinced of this, perturb a teenager.

Control Yourselves!

Control creates order and precision in a system, preventing it from doing random things. Imagine if our neurons fired willy-nilly: we would not be able to get that fork into our mouths, let alone tightrope-walk. Control systems can be geared either for optimal control, which optimizes performance in average, risk-neutral scenarios, or for robust control, which is risk-sensitive and optimizes performance for worst-case scenarios. Optimal control systems, as the name implies, are the best solution to a specific challenge. However, they can be arbitrarily fragile to other uncertainties.[27] Therefore, robust control is what is now primarily used in technological systems (but hidden, and only revealed by what does not usually happen: crashes, stalls, etc.).[28] That Boeing 777 is able to zoom through thunderstorms with flying colors because its control systems are designed to be robust to bad weather, rather than optimized for calm blue skies.

Most neuroscientists have viewed the brain as having *optimal* control systems, even though the neuroscientist, engineer, and physician Daniel Wolpert and his colleagues don't agree. They think, instead, that *robust* control best explains human motor control.[29] Robust control systems have hard limits on robustness and efficiency[30] and must make trade-offs between them: speed versus accuracy; speed versus flexibility; flexibility versus efficiency; speed versus cost. And these trade-offs are well documented in conscious and non-conscious processing of all kinds.[31]

To Wolpert, motor control is the be-all and end-all. One of a line of self-declared motor chauvinists, he boasts a lineage that includes the Nobel laureates Sir Charles Sherrington, who wrote, "Life's aim is an act, not a thought," and Roger Sperry, who encouraged us "to view the brain objectively for what it is, namely, a mechanism for governing motor activity."[32] After all, it is *action*, not cogitation, that puts food on the table and a bun in the oven. Action allowed our ancestors to survive and reproduce. Wolpert, perhaps the current leader of this pack, claims that the only reason we have a brain is so that we can move in an adaptable manner.[33] Before this riles you up, consider that the heart is a muscle whose movement you cannot live without. Motor movement procures food, chews it, and digests it. Without food, the brain can't function and certainly can't produce the creative aspects of life—literature, art, and music—which, in any event, would remain trapped in the brain without motor movement to bring them, through speech, writing, hand gestures, or facial expressions, to the external world. We need to consider this idea with the perspective it gives us. If our brain has evolved as the motor control system for the body, then thinking, planning, remembering, using the senses, and so forth are simply tools, added complexities in a layered architecture that have evolved to increase the robustness of motor control in changing and uncertain environments. That goes for learning and cognition, too. As is always the case, these evolutionarily newer layers bring with them their own fragilities.

If you unexpectedly touch a hot burner, your output will be an automatic reflex: you jerk your finger away before you consciously feel the pain. This is feedback control at the level of the peripheral nervous system. The fast, thick, insulated (and hence costly) spinal cord neurons immediately instigate withdrawal from the painful stimulus, with no help from the conscious brain. The reflex is automatic, fast, costly in energy, and hidden from conscious awareness, but not flexible. You jerk your hand back in one smooth movement; it doesn't sometimes flutter slowly like a butterfly. After a moment's delay, however, the slow, thin, specialized nerves kick in, providing the conscious intel on the source: Ohhhh, my finger hurts. And what happens next? Slow conscious cognition produces a variety of responses about what to do to alleviate the pain now and in the future: You might stick it in your mouth, stick it in a bowl of ice water, or rub aloe vera on it. You plan not to touch a hot burner again. Cognition is accurate, flexible, and energetically cheap, but slow. Sometimes we can afford the time, but in uncertain situations slow may be deadly.

We can think of learning and cognition as tricked-out control layers that have evolved in the attempt to be robust to *future* perturbations by planning for a stimulus that hasn't happened yet. They partially do this by using feedback from previous experiences with a similar stimulus (memory), but this feedback can lead to more than adjusting the stimulus input. It can, over periods of time, make changes to a layer's protocol, which we call learning.

Humans, animals, and some other organisms can prepare for future events through various mechanisms of learning. If an animal encounters a food that tastes bad, it will learn to avoid the stimulus in the future. We know that learning has occurred when the same exact input ("Hot! Don't touch!") produces a different output response in an organism over time. The protocol has changed from "Eat birds" to "Eat birds except crows" or from "Investigate things by touch" to "Investigate things by touch except stoves."

While we are born with some automatic behaviors, like the instinctual pain reflex, some automatic behavior is learned. Learning can shift some behaviors from the tricked-out yet slow conscious control layer down a notch to the fast, automatic nonconscious layer. For example, as you practice your golf swing, you make a prediction from your memory of previous swings where the ball will end up. After you swing, you also get feedback from your vision as to where it actually did end up. Dang. Then you adjust your swing a tad and try again, with a bit more oomph. Good on length but a little wide. Okay, more follow-through on the swing. Enough trials, and the ball ends up right on the money almost every time (as long as there are no external perturbations, such as a gust of wind or your friend cracking a joke just as you swing, or internal perturbations, such as thirst or a strained muscle or a random thought about, well, anything). You no longer need to think about controlling all your movements; they become automatic.

Being robust to future perturbations also involves planning for things that may not have previously been experienced. We plan by using internal models to simulate the future. We recall past events and recombine the memories in a variety of ways to produce a variety of plans to suit possible future circumstances. Thus, the protocol for simulating plans in this tricked-out control layer is a constraint that deconstrains. This is what you did when you reverse-engineered your outfit for the trip north. First, you predicted what you might face and simulated how you would feel in the future from memories of what you had experienced in the past in similar situations. You remembered there was a mismatch between what you wanted to feel and what you did feel, so to adjust that you made your outfit more robust to cold, wet weather. As experiences and the memories they produce accumulate, the planning protocol gains more information to combine into a wider array of future scenarios. The more experiences you have had, the more choices your brain can simulate.

Bringing the idea of layering to a wildly complex biological thing like you and me is really bringing a viewpoint, a stance on how to think about how the gooey biological thing may be working. Breaking matters down into interacting layers gives the engineer a framework for thinking about how to build a brain. While no one is even close to doing so, the perspective does guide the neurobiologists toiling at their benches, studying individual neurons or small circuits of neurons, in how to think about their findings. It suggests how a complex system full of local parts can be organized to get a very large task done, like designing that opera house.

6.

GRAMPS IS DEMENTED
BUT CONSCIOUS

It doesn't make any difference how beautiful your guess is.
It doesn't make any difference how smart you are. . . .
If it disagrees with experiment, it's wrong.

—Richard Feynman

CONSCIOUSNESS IS RESILIENT and hard to stamp out. I learned this when I had the good fortune to spend a few years on the neurology wards. What became apparent from talking to and examining patients with various parts of their brains malfunctioning was that consciousness is truly tough to eliminate. Some form of consciousness always persists except when coma or a vegetative state is brought about by extensive cortical damage that leaves the entire brain dysfunctional. Such damage can be the consequence of leaving a helmet behind when it was most needed, a clot or rupture and bleed of a cerebral artery, the surgical removal of a dangerously situated tumor, or a drug overdose. Otherwise, the personality may change, specific abilities may be lost forever, a person's personal reality may even change, but consciousness keeps coming at you. For sure, the holy grail of science is to find consciousness in the brain, but trust me, it would have been found by now if there were such a thing to find.

Scholars for the past two thousand years of human history have wanted to find the source, the one thing—a spiritual essence, a gland in the forehead, an immortal soul, a brain region—that is responsible for the likes of language, memory, attention, and consciousness. While nobody really knows how any of these capacities we hold so dear actually work, we do know which parts of the brain manage language, memory, and attention. Yet when we try to find the parts of the brain primarily responsible for consciousness, we start to babble, to feel frustrated, because it looks like no such place is to be found. The neurology clinic keeps telling us to try to think about the problem in a different way.

It is always pointed out that there are brainstem lesions that have devastating effects on consciousness, effects so large that people drop into a coma and frequently never come back.* But that is a different kind of thing. If your car was disconnected from its battery, you could never see it work. You couldn't turn it on and see the things it can do. It's similar with lesions to the brainstem: the lion's share of the brain never turns on. There is nothing to observe in extreme cases, and nothing more to understand from them on the topic of consciousness.

Armed with our notions of modules and layers, however, we now can approach the baffling problem of consciousness persisting in the face of various devastating injuries. We have to figure out a way to view how the vast

---

*Processes necessary for consciousness begin in the evolutionarily oldest part of the brain, the brainstem. The main job description for the brainstem is homeostatic regulation of the body and brain. It keeps your heart pumping, your lungs breathing, your guts digesting. Disconnect the brainstem in any mammal, and the body dies. From the brainstem, neurons take off in many directions. Those necessary for consciousness are connected to the intralaminar nuclei (ILN) of the thalamus, which is situated between the midbrain and the cortex.

majority of humans behave when the brain is malfunctioning. We have to grasp why consciousness persists.

Overall, the persistence can be attributed to the multitude of modules that continue to contribute to our daily experiences in spite of injury or other malfunctions. Multi-modular brains have at their beck and call a tremendous number of paths to conscious experience. If one route gets destroyed, another may provide an alternate course. To stamp out consciousness, all modules leading to a conscious state must be shut down. Until this happens, intact modules will continue to pass information from one layer to another and induce a subjective feeling of experience. The contents of that conscious experience may be very different from normal, but consciousness remains. Visiting the neuropsychology clinic, we will see how various assaults on our brain affect consciousness and provide insights into how our brains are organized. It turns out that the endless fluctuations of our cognitive life, which are managed by our cortex, ride on a sea of emotional states, which are constantly being adjusted by our subcortical brain.

Visiting the Clinic

The first patient we meet could be anyone's grandparent. Grandpa shakes my hand in acknowledgment, but he is confused as to who I am. He doesn't remember meeting me a couple of days before. He suffers from the most common type of dementia, Alzheimer's disease, which is associated with the production and accumulation of amyloid-β in the brain. That means he has serious neural damage all over the brain. While for the past twenty years or so amyloid has been considered the "cause" of Alzheimer's, there is recent evidence against that hypothesis, and others are being entertained.[1] At any rate, the disease results in the slow destruction of the brain, commencing particularly with the loss of neurons in the entorhinal cortex and the hippocampus, resulting in short-term memory loss. The disease can become so debilitating that it can completely reshape Grandpa's personality, transforming him from a lively and caring person into a listless shell of his former self. Yet, though he may not recognize me, he is still cognizant of social niceties and shakes my hand. He may wander off, but he will still feel fear when confused and lost, and anger when frustrated. His conscious experience of the world is brought to him through whatever operational neural circuitry continues to function, and as he loses function, it becomes more restricted. The contents of that conscious experience most likely are odd, very different from those of the normal brain or his past self. As a result, odd behavior follows.

The listless version of the formerly jovial grandfather, for example, may still describe himself as his earlier "life of the party" version. Caretakers and family members often attribute a patient's incongruent self-identity to the disorienting nature of the disease. Yet, when friends and family describe the premorbid personality of a loved one, it is strikingly similar to the self-description provided by the individual in the diseased state.[2] This suggests that Grandpa's false beliefs about his current personality traits are likely due to an inability to update those beliefs. Dementia has left Grandpa with an outdated self-image. As long as Grandpa's heart continues to beat, consciousness, albeit with a checkerboard of altered contents, will survive the carnage of his degenerating brain.

Our next visit is to a patient known as Mr. B. He has a different kind of problem. He believes he is of special interest to the FBI, which monitors him every single moment of his day. Not only that, the FBI agents film and broadcast his life to the public as *The Mr. B Show*. Understandably disturbed by this, Mr. B attempts to avoid embarrassing situations by adjusting his behavior. He wears a bathing suit every time he showers, and he changes his clothes under cover of the bedsheets. He avoids social situations, knowing that everyone he encounters is an actor trying to elicit drama to make *The Mr. B Show* more intriguing. One can barely imagine what it would be like to live in Mr. B's world. And yet, when carefully analyzed, Mr. B's case may reveal that a totally rational and normal cortex is trying to make sense out of some abnormalities going on in another region of the brain, the subcortex.

Mr. B suffers from chronic schizophrenia. Risk factors for the disease include a genetic vulnerability and gene-environment interactions. Environmental factors that increase the risk include growing up in urbanized areas,[3] being an immigrant,[4] especially when socially isolated—such as living in an area with few others of the same group[5]—and exposure to cannabis.[6] No matter what evidence is provided to combat Mr. B's false beliefs, he is profoundly convinced that he is constantly being viewed by millions of people. A first-rank symptom of schizophrenia is the perception that particular stimuli, ranked unimportant when in a non-delusional state, are extremely and personally significant:[7] the guy who glances up from his newspaper is deliberately looking at you; the rock on the road was deliberately placed to harm you. This alteration in salience, that is, what is important and draws one's attention, is such a classic feature of schizophrenia spectrum disorders that there is a growing movement pushing for the tag "schizophrenia" to be abandoned and the disorder reclassified as a "salience syndrome."[8]

A sensory input becomes more salient when the neural signal that it elicits is enhanced over others, which draws attention to it. Shitij Kapur, a psychiatrist, neuroscientist, and professor at King's College London, distinguishes for us the difference between hallucinations and delusions: "Hallucinations reflect a direct experience of the aberrant salience of internal representations," whereas delusions (false beliefs) are the result of "a cognitive effort by the patient to make sense of these aberrantly salient experiences."[9] In the brain, the amount of the neurotransmitter dopamine affects the process of salience acquisition and expression. During an acute psychotic state, schizophrenia is associated with an increase in dopamine synthesis, dopamine release, and resting-state synaptic dopamine concentrations.[10] Kapur suggests that in psychosis, there is a malfunction in the regulation of dopamine, causing abnormal firing of the dopamine system, leading to the aberrant levels of the neurotransmitter and, thus, aberrant assignment of motivational salience to objects, people, and actions.[11] Research supports this claim.[12] The altered salience of sensory stimuli results in a conscious experience with very different contents than would normally be there, yet those contents are what constitute Mr. B's reality and provide the experiences that his cognition must make sense of. When considering the contents of Mr. B's conscious experience, his hallucinations, his efforts to make sense of his delusions are no longer so wacky, but are possible, though not probable, explanations of what he is experiencing. With this in mind, the behavior that results from his cognitive conclusion seems somewhat more rational. And despite suffering this altered brain function, Mr. B continues to be conscious and aware of his existence.

The Walking Unconscious

Strange behaviors, however, can also arise from a fully intact and functioning brain if only part of it is awake. In a layered brain, lots of activities are happening simultaneously and are coordinated synchronously. What if things get out of sync—if every layer is working, but out of step? Our next visit is to Mr. A, perhaps the most unsettling of our visits.

Mr. A, described by family and friends as a loving family man, was awakened in his bed by his dogs' barking and strange voices. Racing downstairs, he was met by several police officers with their guns drawn.[13] Dazed and confused, Mr. A was cuffed and locked in the back of a squad car, trembling in fear as he tried to assess the situation by listening to the conversations of emergency personnel through the window. He gathered that his wife had been badly hurt and thought that the cops were on the hunt for the person responsible. He didn't know until later that they had already found their man, and it was he.

In a frazzled panic, Mr. A could only remember falling asleep in his bed a few hours earlier. The police elucidated the tragic situation. Mr. A had brutally murdered his wife during what was later determined to be a sleepwalking episode. During this episode, he had gotten up from bed and gone out to fix the pool's filter, which his wife had asked him to do at dinner. She must have awoken and gone down to coax him back to bed. His concentration on the motor interrupted, he had turned violent and stabbed her forty-five times, put his tools away in the garage, returned to find her still alive, and rolled her into the pool, where she drowned. Then he returned to bed. His neighbor, hearing screaming and barking next door, looked over the fence to see a "bewildered"-appearing Mr. A roll a body into the pool, and called the police.

The idea that someone could kill his wife, whom he loved, while sleep-walking is unfathomable. Yet, with no identifiable motive, no attempt to hide the body or weapon, and no memory of the event, the jury was convinced that his actions occurred unintentionally and out of Mr. A's awareness. If this is true, what exactly went on in Mr. A's mind and brain during this atrocity?

Sleepwalking is a *parasomnia*, a strange behavior that occurs during sleep. Over the years, sleep experts have identified two main stages of sleep by recording brain waves—rapid eye movement (REM) and non–rapid eye movement (non-REM) sleep. Sleepwalking usually occurs after an abrupt and incomplete spontaneous arousal from the non-REM sleep that occurs in the first couple of hours of the night, turning one into a mobile sleeper. Trying to waken sleepwalkers is fruitless and can be dangerous, since the sleepwalker may feel threatened by physical contact and respond violently. Normally, non-REM sleep shifts into REM sleep, during which there is a loss of muscle tone, preventing motor behavior during REM sleep. The majority of sleep-walking episodes tend to be relatively harmless and usually make for a good story as told by the witness, often beginning with "You won't believe what you did last night!" And if you are the sleepwalker you don't believe it, because you will have no memory of your midnight shenanigans.

Most parasomniac behaviors appear irrational and are disconcerting to watch. The sleepwalker may start vacuuming or sweeping the patio in the middle of the night, oblivious to surroundings. In rare instances, sleepwalkers engage in very complex and potentially dangerous activities, such as mowing the lawn, repairing a motorcycle, or driving a car. Complex behaviors such as these make it difficult to believe that the sleepwalker is not consciously aware of his or her actions in the moment. Rarely, these complex behaviors turn violent. When the law is involved, whether the behavior was intentional or not becomes key, thus intensifying the debate over whether or not these people are conscious during their sleepwalking activities.

A clearer picture of what is happening in the brain during non-REM sleep,[14] during sleepwalking,[15] and during confused arousals[16] has been achieved through neuroimaging and EEG. It appears that the brain is half awake and half asleep: the cerebellum and brainstem are active, while the cerebrum and cerebral cortex have minimal activity. The pathways involved with control of complex motor behavior and emotion generation are buzzing, while those pathways projecting to the frontal lobe, involved in planning, attention, judgment, emotional face recognition, and emotional regulation are zoned out. Sleepwalkers don't remember their escapades, nor can they be awakened by noise or shouts, because the parts of the cortex that contribute to sensory processing and the formation of new memories are snoozing, temporarily turned off, disconnected, and not contributing any input to the flow of consciousness.

It is likely that Mr. A consciously experienced aspects of the episode, but very differently than his awake self would have. With a layered brain in mind, we can predict that certain "lower-level" consciousness-producing modules were active, allowing him to adeptly navigate and coordinate movements and feel emotions, while other, "higher-level" ones remained asleep and silent, preventing him from comprehending the situation, recognizing his wife, hearing her screams, or remembering the event. System-wide, particular regions were isolated and disconnected, and only certain modules were contributing to his behavior and his conscious experience. Unfortunately, the cortex was asleep, locked up tight and contributing nothing. When Mr. A snapped out of his sleep cycle, these silent modules awoke to a nightmarish reality. During this horrendous incident the awake regions of his undamaged brain, unfettered by the processing of the cognitive control modules of his sleeping cortex, produced behavior that markedly strayed from that of this normally compassionate, nonviolent man. The fact that these actions completely went against Mr. A's personality traits and ideals is exactly why jurors settled on a "not guilty" verdict.

Unmoving but Conscious

In contrast, one of the most nightmare-inducing brain injuries is a lesion to the ventral part of the pons in the brainstem. The loss of these neurons, which connect the cerebellum with the cortex, leave one unable to move but fully conscious. This famously happened to Jean-Dominique Bauby, the editor in chief of the French *Elle* magazine, when he suffered a stroke at the age of forty-three. Waking up several weeks later from a coma, fully conscious and with no cognitive loss, he was unable to move anything except his left eyelid.[17] That also meant he couldn't talk, and thus couldn't tell anyone that he was conscious. He had to wait until someone noticed that he appeared to voluntarily blink his eyelid. This is known as "locked-in" syndrome. The lucky ones, if you can call it that, can voluntarily blink or move their eyes, though the movement is small and tiring. This is how they communicate. The unlucky ones cannot.

In many cases, it has taken months or years before it was recognized by a caregiver that the patient was conscious, suffering medical procedures without anesthesia and hearing conversations about his own fate that he could not participate in. Once it was recognized that Bauby was conscious, he took advantage of his ability to blink his eye. He wrote a book describing his conscious experience as he lay paralyzed. He would construct and memorize sentences as he lay there. Then, for four hours a day, an amanuensis patiently sat at his bedside going through a frequency-ordered French alphabet, and Bauby would blink when the correct letter was spoken. Two hundred thousand blinks later, *The Diving Bell and the Butterfly* was done. In the prologue, speaking in the third person, he describes the condition he awoke to find himself in: "Paralyzed from head to toe, the patient, his mind intact, is imprisoned inside his own body, unable to speak or move. In my case, blinking my left eyelid is my only means of communication."[18] He describes feeling stiff and being able to feel pain, yet he goes on to say,

My mind takes flight like a butterfly. There is so much to do. You can wander off in space or in time, set out for Tierra del Fuego or for King Midas's court.

You can visit the woman you love, slide down beside her and stroke her still-sleeping face. You can build castles in Spain, steal the Golden Fleece, discover Atlantis, realize your childhood dreams and adult ambitions.[19]

Bauby is an example of the endless capacity of human adaptability. In fact, adaptability appears to be de rigueur for such patients, for 75 percent have rarely or never had suicidal thoughts.[20] Even with this devastating injury to part of the brainstem, consciousness remains, accompanied by the full range of feelings about both present and past experiences.

When modules and layers are damaged or malfunctioning, strange behaviors can result. From the widespread cortical damage and disruption in Alzheimer's disease to the specific disorders of brainstem damage, a picture begins to emerge: it takes an understanding of both the cortex and the sub-cortex to capture the ever-changing moments of conscious experience. Could it be that all of those fleeting conscious thoughts occur on a bed of a few specific emotional states that give those thoughts a subjective feel? Could it be this all fits into the brain's layered architecture, with the evolutionarily older brain system—still wired up to signal the organism to fight or run, or to seek mates or to eat—operating outside the direct control of the cognitive layers? Is the layered-architecture model going to give us the means to comprehend how we are organized to be conscious?

Subcortical Emotional Feelings Engine

A long-standing belief is that the cerebral cortex is responsible for all forms of consciousness, and that without it we would be not just unconscious but with no capacity to be conscious at any level, that is, a conscious-less being in a vegetative state.[21] The cortex, however, could simply be a collection of extensions (apps!) to enhance conscious experiences. Sure, it provides us with several dynamic mental skill sets—that is, ones that can change and are constantly active—but it might not be essential for giving us a raw subjective feeling. Underneath the cortical hood are several subcortical networks that are essential for maintaining consciousness. It is lesions to these subcortical regions that can result in a coma, where a person or animal becomes nonresponsive and appears unconscious to an outside observer.[22] Even a fully functional cerebral cortex cannot salvage the wreckage of some types of subcortical damage.

The challenge of drawing the line between a conscious and an unconscious state has, in the past, largely been semantic. The term "consciousness" lacks objectivity because it is difficult to define a subjective feeling of existence. This is the main reason why the defining features of consciousness are hotly debated. As soon as one steps into the clinic, however, identifying conscious states is of the utmost urgency, and is no longer simply a semantic problem but an ethical one, too. Denying pain medication to a seemingly unconscious patient who, unbeknownst to you, is conscious is torture. Despite the ambiguity surrounding the term, there is compelling evidence to suggest that the cerebral cortex is not necessary to evoke some forms of consciousness. The capabilities of subcortical systems appear competent enough on their own to provide a subjective feeling.

That evidence comes from the pediatric clinic. Sadly, some children are born with anencephaly (without a cerebral cortex due to genetic or developmental causes) or hydranencephaly (very minimal cerebral cortex, often the result of fetal trauma or disease). The neuroscientist Björn Merker became interested in the subcortex early in his career. Frustrated by the limited information on and few case studies of children with hydranencephaly, he joined a worldwide Internet group of parents and caretakers of these children to learn more about them and their condition. He came to know several families and spent a week with them at Disney World. Over the course of that week he observed that the children "are not only awake and often alert, but show responsiveness to their surroundings in the form of emotional or orienting reactions to environmental events. . . . They express pleasure by smiling and laughter, and aversion by 'fussing,' arching of the back and crying (in many gradations), their faces being animated by these emotional states. A familiar adult can employ this responsiveness to build up play sequences predictably progressing from smiling, through giggling, to laughter and great excitement on the part of the child."[23] Without a cerebral cortex or the cognition it supplies, these children were feeling emotions, having a subjective experience, and were conscious. No one would mistake them for a child with a cerebral cortex, but they are aware and their emotional response to stimuli is appropriate.

Over the years, Merker has reached the conclusion that it is the midbrain that supports the basic capacity for conscious subjective experience. Sure, the cortex elaborates on the contents of the experience, but the capacity itself arises from the midbrain structures. The ethical implications of this are obvious. Merker notes that parents often encounter medical professionals who are surprised when asked for pain medication for these children when they are to undergo invasive procedures.

The main argument opposing the idea that these children are experiencing the world through their subcortical structures is drawn from the fact that almost all of them have some portions of their cerebral cortex spared. Yet although the very limited intact cortical regions (of questionable functionality) vary widely from child to child, their behavior is fairly consistent, and it is asymmetrical with the tissue that is present. For example, while the auditory cortical tissue is rarely preserved, hearing is usually preserved, and while it is common for some visual cortex to be preserved, vision tends to be compromised.

Merker's theory is bolstered by research into the emotional life of animals. The Estonian-born neuroscientist Jaak Panksepp studied the nature of emotions in animals for half a century. He differentiated two types of consciousness: the evolutionarily old affective consciousness (conscious of raw emotional feelings) and the relative newcomer, cognitive consciousness (which allows one to think about those emotional feelings). In a lecture, he told the story about the final lab practicum he gave to his undergraduates at the end of the course. He would prepare two rats for each student to study. One of the rats was decorticated, leaving only subcortical tissue. The other rat was given a sham operation, which means it underwent surgery, but nothing was actually removed from the brain. The students were to study their pair of rats for two hours on a wide range of tasks that they had learned about in the class. When time was up, they had to guess which rat had lost all of its cerebral cortex and explain their choice. Twelve of the sixteen class members declared the decorticated rats were the neurologically normal ones!

What the students observed in these rats was motivated behavior, such as searching for food, mating, fighting or escaping when attacked, and wrestling playfully with other rats.[24] To those students, the behavior that they observed was ratty enough to declare them normal! If the cortex is solely responsible for mediating consciousness, then removal of the cortex should have caused those rats to be unresponsive to other playful rats and everything else. Yet removing their cortex did not stamp out their basic competences and responses, meaning that the upper brainstem mechanisms were enough to sustain many of their behaviors, including their emotional and motivational feelings, discussed below.

Feelings Tie to Consciousness

These studies of decorticated rats and children with hydranencephaly suggest that subcortical structures can transform raw neural input into something resembling core emotional feelings. Subcortical brain areas have their own dynamics, arose early in the evolutionary process, and are anatomically, neurochemically, and functionally homologous in all mammals that have been studied.[25] Panksepp argued that we share these brain areas that produce the emotions we feel with other animals, and these areas have been selected for their ability to enhance survival. How? The emotions act as an interior reward and punishment system that informs an animal how it is far-ing in the quest for survival. Positive emotional feelings egg the animal on, while negative feelings, depending on their strength, indicate anything from iffy to disastrous situations. Thus, these internal senses provide a way to assess the external environment and are powerful drivers of behavior, despite their relative simplicity at the level of consciousness.

The Caltech researchers David Anderson and Ralph Adolphs are on the same page as Panksepp.[26] They argue that an emotion is an unconscious central nervous system state that is triggered by a specific stimulus, whether it be external, such as a predator, or internal, such as the memory of one. When activated, the neural circuit that encodes this state causes multiple parallel processes to kick into gear, which produce a behavioral response, feelings, cognitive changes, and somatic responses, such as a racing heart and a dry mouth. One can feel without cognition kicking in and reporting that feeling.

According to Panksepp, seven primal emotional and motivational feel-ings that appear to be common features of animal and human consciousness at both a behavioral and a neural level are SEEKING, FEAR, RAGE, LUST, CARE, GRIEF, and PLAY. These feelings, which can largely be attributed to functions of the subcortical limbic system, drive animals to behave in ways that promote finding food, shelter, and mates; avoiding harm; protecting one-self and kin; and building relationships with friends and family. If we consider consciousness to be a subjective feeling about something, then we must consider emotions to be a foundational component of consciousness.

Panksepp concluded that emotional feelings were such a successful tool for living that they were coded into the genome in rough form, have been conserved across all mammals, and only later in the evolutionary process were they gilded with learning mechanisms and higher-order cognitions provided by an add-on extension: the cortex.[27] If these feelings existed before cortical tissues, then the special wiring of these subcortical networks alone must possess what is necessary to produce the feelings that accom-pany conscious experience. By understanding the layered system of subcortical networks, we can, perhaps, better appreciate the most primitive form of consciousness. The emotional and motivational feelings and the behaviors they produce in animals can teach us a lot about how modular systems promote consciousness and perhaps indicate what aspects of human consciousness are unique.[28]

New York University's Joseph LeDoux, who has painstakingly elucidated what he formerly referred to as the fear circuits and now calls threat circuits, has a different take. He has two major concerns. The first is that there still is no agreed-upon definition of emotion; the second is that there are those who challenge the notion that there are common basic emotions. Thus, how can one confidently differentiate emotion from other psychological states or compare emotions across species? LeDoux writes: "The short answer is that we fake it. Introspections from personal subjective experiences tell us that some mental states have a certain 'feeling' associated with them and others do not." This leads him to worry about claims that similar behavior in animals indicates a similarity of experience.[29] From his perspective, the cortex is necessary for affective feelings. He thinks that subcortical circuits produce emotional behavior and physiological responses, but that they only indirectly contribute to subjective feelings. He requires an additional cognitive step for the production of subjective feelings, which is provided by higher cortical circuits that read out and interpret the emotional behavior. He is not alone in holding this view. In fact, the majority of emotion researchers would give such cognitive "readout" theories the thumbs-up. LeDoux suggests that conscious feelings are a two-step process and result when a physiological response is read out by the parts of the prefrontal cortex that support working memory.

While this battle over emotions is fought, we can remain on the sidelines because a layered brain architecture can accommodate either scenario. What is important is that both the subcortex and the cortex contribute to the full conscious experience. From one perspective, the children with hydranencephaly have emotional feelings that appear identical to those of children with intact cortices. Because their overt behavior is similar, we quickly map the full complement of a meta-self-aware (aware that they are aware) conscious experience onto them. Are they self-aware? Without a cortex supplying the necessary functions for cognition, they are unable to know they are self-aware. At a minimum, for a full-blown awareness that you are having a conscious experience, both layers must be functioning.

Consciousness Enhanced by the Cortex

Is the cortex overrated if subcortical circuitry contains the essential ingredients for consciousness? *Mais non!* The point is that the subcortex should not be underrated. By understanding the contribution of subcortical processing to consciousness, we are better equipped to realize why it is so hard to get rid of this incessant feeling of feelings. The cerebral cortex clearly plays the role of providing the contents of consciousness, given how brain damage to that area often coincides with specific behavioral changes. What exactly is the role of the cerebral cortex in producing consciousness? The cortex expands the number of ways in which we can experience the world, which allows for a vast variety of possible conscious experiences and responses.

The particular brand of cortex that each species possesses provides it with its own particular contents of conscious experience. Part of the contents of human consciousness is language. Only humans have come up with nifty little symbols that, in a specific combination, can give another person a specific mental representation of some abstract idea. Not only do we have the capacity to learn language, but we are also biologically prepared for language acquisition.[30] As we discussed in chapter 4, we have entire brain regions dedicated to various aspects of learning, comprehending, and producing language. Another trip to the clinic will show us that damage to one region will destroy our ability to comprehend words, but leave us able to produce grammatically correct nonsensical sentences with proper prosody and intonation. Lesions in a different area will have us comprehending sentences but unable to construct them. Lesions in another area, and you will be unable to say nouns but still be capable of recognizing and comprehending them. Any such damage will result in a different conscious experience, but none will destroy consciousness itself.

While language adds to our conscious experiences, we would still be conscious without it, though many of our experiences would be markedly different. Consider the life of the French feral child Victor of Aveyron, immortalized in François Truffaut's 1970 film *L'Enfant Sauvage*. Found at age twelve, Victor had spent his childhood alone in the woods, never having been exposed to language, and had never learned to speak. He was definitely conscious and having conscious experiences, yet with contents different from what they would have been had he learned to speak. When the functionality of one module does not develop, other modules step in to give you an alternative experience.

There is a lot of debate about whether subcortical structures are the main driving force of consciousness[31] or whether consciousness is primarily mediated by the cerebral cortex.[32] When thinking about brain functioning, however, there may not be a specific modular hierarchy that allows consciousness to manifest itself in one way or another. Specific modules work relatively independently and, rather than being a neatly ordered queue of modular processing, the contents of our conscious experience may be the result of some kind of competition: some processing takes hold of your conscious landscape at a given moment in time, and some doesn't. In this view, both subcortical and cortical modules have the capacity to produce a form of conscious experience that does not necessarily require intervention from "lower" or "higher" cognitive systems. Rather, the multitude of conscious-producing modules simply diversifies your conscious portfolio. To better illustrate this concept, let's try to stamp out consciousness with an iron rod.

One of the most fascinating and famous brain lesions in history arose from a railway construction explosion that sent a searing hot metal rod through the skull and left frontal lobe of a construction worker named Phineas Gage. Surprisingly, Phineas did not appear to lose consciousness even moments after the accident! Personally, I'd rather have experienced Muhammad Ali's anchor punch that knocked out Sonny Liston than have an iron rod bust open my cranium, but apparently Ali's punch more effectively stamped out consciousness (at least momentarily). Although Sonny recovered from Ali's punch, immediate and permanent brain damage ensued for Phineas Gage. Despite missing half of his frontal lobe, Phineas could still function similarly to the way he did before his accident. His actual manners drastically changed, however, as the once professional and respectful Phineas became a lewd, disrespectful man.[33] Phineas became less conscientious in his actions toward others, but he did not become any less conscious. His range of possible conscious experiences diminished a bit, as it seemed his once sympathetic attitudes toward colleagues were replaced with experiences of agitation and aggression. Phineas Gage suffered from what is now called *frontal lobe syndrome*, in which he lost all functionality of his left frontal lobe. When frontal lobe damage occurs, people tend to have difficulty regulating their emotions.[34] This loss of emotional control might be attributed to the subcortical modules "winning" the competition to provide an overarching conscious experience more often, since there is less competition from modulating frontal tissues. Regardless of the underlying reason for loss of emotional control in frontal lobe syndrome, one fact remains consistent across all cases: the person is still conscious.

There are a tremendous number of brain lesion cases that paint a similar picture: Damage or dysfunction in brain region X causes a change in behavior Y, but consciousness almost always remains intact. The modular brain makes consciousness resilient because of the plethora of possible paths that can lead to a conscious moment. Only a brain organized in this way can explain these facts of neurology. Losing modules causes losses in specific functionalities, but the mind keeps on producing a continuous conscious stream as if nothing changed. The only thing that has changed is the contents of that stream. Not only does this provide evidence that the brain operates in a modular fashion, but it also suggests that independent modules can each produce a unique form of consciousness.

The Ubiquity of Consciousness

What we have learned from visiting the neurology clinic is that severe brain damage across various locations of the brain cannot stamp out consciousness per se. Certain contents of conscious experience may be lost, but not consciousness itself. This fact suggests that there is not a specific "Grand Central" cortical circuit that produces consciousness, but that any part of the cortex can produce it when supported by subcortical processing, and that subcortical processing alone can support a limited type of conscious experience. Thus, it appears that it is the processing of local modular circuits that provides the contents of conscious experience.

Although these modular systems are largely independent, communication between modules helps coordinate the flow of consciousness. This communication is important to keep each module up-to-date on recent personal events. Just as the news informs citizens of worldly events as they occur, the connections between modules coordinate information to make sure all modules are functioning on the same page. We only notice when communication is a day late and a dollar short. Rustling noises outside the back door at night may activate the subcortical fight-or-flight response, and you grab your phone to call the police, only to realize in the next moment when cognition kicks in that it is a raccoon going through your trash, again. Thank your limbic system for quickly getting you prepared for a potentially dangerous situation, but there is no danger here, just another mess to deal with in the morning.

This incessant interplay between cognition and feelings, which is to say between cortical and subcortical modules, produces what we call consciousness. There obviously is a different feel to a wave of intense emotion versus an abstract thought, but each conscious form is an experience that gives us a unique perception of reality. The pattern in which these various conscious forms come in and out of awareness gives us our own personal life story. The vast variety of conscious forms and the ubiquity of consciousness in the brain are best explained by a modular architecture of the brain. The conceptual challenge now is to understand how hundreds, if not thousands, of modules, embedded in a layered architecture—each layer of which can produce a form of consciousness—give us a single, unified life experience at any given moment that seems to flow flawlessly into the next across time. The key idea, as we shall see in chapter 9, is time. It is the unending sequence of modules having their moment.

How that works is coming up, but before getting to it, we have to deal with the elephant in the room. Whatever model one has for how the brain does its trick of turning neuronal firings into mental events, we must try to understand the gap between those two phenomena, one objective (neuronal) and the other subjective (mental), and whether bridging that gap is even possible. No matter whether you think that local modules are responsible or that central brain circuits underlie what we call conscious states, you still have to deal with the gap. Some feel this is an impossible assignment.

To get at this fundamental question, we're going to have to look back at what the mathematicians and physicists have been thinking about for the past 150 years. After all, their lot came up with perhaps the greatest ideas in human history, the theory of relativity and quantum theory. Their thinking was truly on the edge of human mental capacity, and they were grappling with phenomenal unknowns. The fruits of their thinking were virtually ignored by biologists, psychologists, and neuroscientists and dismissed as irrelevant to the problem of consciousness. I think they can help, because what has gone underappreciated from math and physics is the idea of *complementarity*, which holds that a single thing can have two kinds of description and reality. Could that idea help us with the deep divide between the mind and brain? Could it help us understand the "explanatory gap" between the reality of the physical world—that material brain of ours made up of chemicals governed by the laws of physics—and the reality of that seemingly nonmaterial subjective experience? I think it can, and before we get ahead of ourselves, let's familiarize ourselves with the physics that may yet hold the key.

PART III:

CONSCIOUSNESS

COMES

7.

THE CONCEPT OF COMPLEMENTARITY: THE GIFT FROM PHYSICS

Those who are not shocked when they first come across
quantum theory cannot possibly have understood it.

—Neils Bohr

I N 1868, THE physicist, mountaineer, educator, and professor at the Royal Institution John Tyndall gave a talk to the mathematical and physical section of the British Association for the Advancement of Science. In it, he laid out the following dilemma:

> The passage from the physics of the brain to the corresponding facts
> of consciousness is unthinkable. Granted that a definite thought,
> and a definite molecular action in the brain occur simultaneously;
> we do not possess the intellectual organ, nor apparently any rudiment
> of the organ, which would enable us to pass, by a process of reasoning,

from the one to the other. . . . "How are these physical processes connected with the facts of consciousness?" The chasm between the two classes of phenomena would still remain intellectually impassable.[1]

Here we are, 150 years later, and we are not quite as far as we would like to be. We understand, to some extent, the electric discharges, the groupings and flow of molecules, and sometimes even corresponding brain states, especially in the study of vision. Unlike Tyndall, however, I think we do have an organ that is up to the task. What is needed is to apply the right kinds of ideas to the problem of determining how mind comes from brain. How do we think about that pesky gap between our biology and our mind?

It is commonly recognized that the chasm or gap is a problem. It was only twenty-five years ago that the philosopher Joseph Levine officially dubbed it the *explanatory gap*, which he later described in his book *Purple Haze*:

> We have no idea, I contend, how a physical object could constitute a subject of experience, enjoying, not merely instantiating, states with all sorts of qualitative character. As I now look at my red diskette case, I'm having a visual experience that is reddish in character. Light of a particular composition is bouncing off the diskette case and stimulating my retina in a particular way. That retinal stimulation now causes further impulses down the optic nerve, eventually causing various neural events in the visual cortex. Where in all of this can we see the events that explain my having a reddish experience? There seems to be no discernible connection between the physical description and the mental one, and thus no explanation of the latter in terms of the former.[2]

Levine leaves us with an unbridged chasm between the physical level of interacting neurons and the seemingly nebulous level of conscious experience. We may explain, for example, that pain is caused by the nervous system's firing of C fibers, and why there is a delay between withdrawal and the feeling of pain, but explaining the causal relationship tells us nothing about the feeling of pain itself, that subjective experience.

The current state of the mind/body problem rests on two plausible yet seemingly incompatible propositions: (1) Some form of materialism or physicalism is true. (2) Physicalism cannot explain phenomenal consciousness, raw feel, or qualia. Pick (1) and you are a materialist; pick (2) and you are a dualist. Levine, however, throws caution to the wind and picks both. He is a materialist *and* believes that phenomenal facts can never be derived from physical facts. Can he have his cake and eat it, too? Most philosophers and neuroscientists would say he cannot, so how does he do it?

Levine junks the intuition that mental events (those much heralded qualitative experiences) seem different from physical events. For example, the delay in feeling pain is a phenomenal fact explained by a physical fact. But that is not Levine's issue. While he accepts that firings of neurons cause phenomenal experience, and that consciousness really must be a physical phenomenon, he states, "There are two, interrelated features of conscious experience that both resist explanatory reduction to the physical: subjectivity and qualitative character." If we can't bridge that gap by explaining how the firing of neurons = the experience of pain, then Levine suggests that "it must be that the terms flanking the identity sign themselves represent distinct things."[3] That surprisingly sounds like Levine has resorted to a form of dualism.

Several years later, however, Levine made clear that he did not believe in an actual gap, a nothingness between neurons and subjective experience. He simply was pointing out we have no knowledge about how that gap might be closed. Of course, when you think about it, gaps are all over the place in the history of science, but they are usually framed in terms of gaps in knowledge. In the end, Levine thought that this is also the case for the mind/brain gap. In fancy philosophical terms, he believed it is a question of epistemology versus metaphysics. He viewed it as a gap in current understanding as to how such things are to be explained. Of course, when put that way, he is completely correct.

An even stronger view comes from the Australian philosopher David Chalmers, who also agrees there is an "explanatory gap." He is steadfastly committed to the perspective of proposition (2): Physicalism cannot explain phenomenal consciousness, raw feel, or qualia. This makes him a dualist, though Chalmers would specify that he is a naturalistic dualist. He agrees that mental states are caused by the physical systems of the brain (that's the naturalist part), but he believes that mental states are fundamentally distinct from and not reducible to physical systems.[4] This is an extraordinary position for a modern philosopher, but not for non-philosophers. Most people on earth today are dualists!

Yet Tyndall, in 1879, slightly rewording his inaugural address as new president of the British Association given in 1874, and foreshadowing our discussion later in this chapter concerning the origins of life, wrote: "Believing as I do in the continuity of nature, I cannot stop abruptly where our microscopes cease to be of use. Here the vision of the mind authoritatively supplements the vision of the eye. By an intellectual necessity I cross the boundary of the experimental evidence, and discern in that 'matter' . . . the promise and potency of all terrestrial life."[5] William James was of the same opinion; he stated:

> The demand for continuity has, over large tracts of science, proved itself to possess true prophetic power. We ought therefore ourselves sincerely to try every possible mode of conceiving the dawn of consciousness so that it may *not* appear equivalent to the irruption into the universe of a new nature, non-existent until then.[6]

He also said:

> The point which as evolutionists we are bound to hold fast to is that all the new forms of being that make their appearance are really nothing more than results of the redistribution of the original and unchanging materials. The self-same atoms which, chaotically

dispersed, made the nebula, now, jammed and temporarily caught in peculiar positions, form our brains; and the "evolution" of the brains, if understood, would be simply the account of how the atoms came to be so caught and jammed. In this story no new *natures*, no factors not present at the beginning, are introduced at any later stage.[7]

It seems that over the past few decades most of us have forgotten that human consciousness has gradually evolved from precursors; it did not spring fully formed into the brain of the first *Homo whateverensis*. James goes on to comment, *"If evolution is to work smoothly, consciousness in some shape must have been present at the very origin of things."*[8] So, yes, that far down. If we want to get at an understanding of the chasm between mind and brain, we have to delve deeply into other big questions, such as how it is that life comes out of non-living matter.

This chapter's journey will find us discovering that in order to understand what the difference is between living and non-living matter, it is necessary to grasp the inherent duality of all evolvable entities—the fact that, indeed, all living matter can be in two different states at the same time. As you will see, physics and biosemiotics can show us how to resolve the inherent gaps between living and non-living systems without resorting to spooks in the system. The insights of these disciplines suggest how to think about the problem of such gaps in general and how to think about closing this one in particular, and offer a road map for how neuroscientists might succeed with a mind/brain gap nested in a layered architecture, with protocols that describe the interfaces between those layers. But first, the physics.

The Beginnings of Physics and the Commitment to Determinism

The story starts with Isaac Newton and the spectacular beginnings of classical physics in the seventeenth century. This is the kind of physics most of us struggled to learn in school. It turns out that the apple story is the real deal.

Newton himself related it to his biographer, William Stukeley, reminiscing about a day in 1666 when, sitting under an apple tree, he wondered,

> Why should that apple always descend perpendicularly to the ground. . . . Why should it not go sideways, or upwards? but constantly to the earths center? assuredly, the reason is, that the earth draws it. there must be a drawing power in matter. & the sum of the drawing power in the matter of the earth must be in the earths center, not in any side of the earth. therefore dos this apple fall perpendicularly, or toward the center. if matter thus draws matter; it must be in proportion of its quantity. therefore the apple draws the earth, as well as the earth draws the apple.[9]

Newton's niece's husband, John Conduitt, related how Newton went on to wonder if this power might extend beyond the Earth: "Why not as high as the Moon said he to himself & if so, that must influence her motion & perhaps retain her in her orbit, whereupon he fell a calculating what would be the effect of that supposition."[10] Calculate he did. Newton converted the results of Galileo's "terrestrial" motion experiments into algebraic equations, now known as the laws of motion. Galileo had shown that objects retain their velocity and trajectories unless a force acts upon them; objects have a natural resistance to changes in motion, known as inertia; and, finally, friction is a force. That last finding is presented in the third law: To every action there is always an equal and opposite reaction. Newton's apple musings and his various calculations led him to the universal law of gravitation, and to the realization that the "terrestrial" laws of motion that he had put into algebra also described observations that Johannes Kepler had made about the motions of the planets. That is not a bad day's work.

Then came Newton's big revelation. He had just come up with a set of fixed, knowable mathematical relationships that described, well, the workings of all the physical matter in the universe, from bocce balls to planets. These laws are universal, inexorable. They are separate from him, Newton the observer, Kepler, and everyone else. The universe and all the systems it contains just hum along following these laws concerned with space, time, matter, and energy, with or without observers. When the tree falls in the forest with no one to observe it, it still makes sound waves. Whether they are heard or not is another matter, and we shall soon see that the distinction illustrates the crux of our problem about the origins of life.

Newton stirred up more than scientific interest. It was thought that if his laws are universal, then, theoretically, if the initial conditions were known, every action in the physical universe is predictable. That means that all actions are determined, even *your* actions, for you are just another physical thing in the universe. Plug the right initial conditions into the equation, and out will come the answer as to what happens next—even what you are going to do after work next Thursday. But this line of thinking overlooks a crucial point. What we are going to find out in a bit is that plugging in a value for those initial conditions is a subjective choice made by the experimenter, and that subjective choice is a wolf of a problem dressed up in sheep's clothing. It is not so simple.

Newton's laws seem to undermine free will and, thus, responsibility for one's actions. Determinism first captured the imagination of the physicists, and soon many others got caught in its sway. Still, even though Newton's view of things took some getting used to, his laws seemed to describe most observations of the physical world well, and they became entrenched over the next two hundred years. But soon there was a new challenge to Newtonian physics that had to do with a new invention: the steam engine. The first commercial one was patented by Thomas Savery, a military engineer, in 1698 to pump water out of flooded coal mines. Even as the engines' design improved, one problem continued to plague them: the amount of work they produced was minuscule compared to the amount of wood that had to be burned to produce it.

The early engines were all super inefficient because way too much energy was dissipated or lost. In the wholly determined world that Newton envisioned, this didn't make much sense, so the theoretical physicists were forced to confront the puzzle of the seemingly lost energy. Soon a new field of study emerged, thermodynamics, and with it a change in theory about the nature of the world. It all has to do with the relations of heat and temperature to energy and work. When it was all thought through, the field of physics was changed forever, and Newton's determined world looked a bit different.

The Rise of Quantum Mechanics and a Statistical View of Causation

It wasn't long before the steam engine problem gave rise to the first two laws of thermodynamics. The first states: The internal energy of an isolated system is constant. In essence this is reiterating the law of conservation of energy, which states that while energy can be transformed from one form to another, it cannot be created or destroyed. This is entirely consistent with the deterministic world of Newton, but it was also a very limited claim, since it was true only for isolated and contained systems.

The second law is where things get interesting and challenging, and it involves something called entropy. The second law reveals that such things as heat cannot spontaneously flow from a colder to a hotter location. I can remember the moment when I struggled to grasp this concept. It was a cold winter's day at Dartmouth and I was welcoming a physicist to a meeting in my office. He had just walked across the campus Green, a wide-open and cold walk that had nearly frozen his parka itself. I happily remarked that every time someone walked into my office their clothes brought in the cold, and I always felt chilled. He looked at me and said, "Let's get the physics of this straight. Cold is not transferring to you. Your body heat is transferring to me, and because that heat is leaving your body, you feel colder." He reminded me that the second law of thermodynamics can come in very handy even in understanding everyday life, and then added that we needed to hire another theoretical physicist.

"Entropy" was a term originally coined by the nineteenth-century German physicist Rudolf Clausius to describe "waste heat." It is a measure of the amount of thermal energy that cannot be used for work. The physicist's cold parka had increased my state of entropy, and with that, there was less energy available to keep me warm.[11] The second law is where things started getting fuzzy. In short, with parkas and steam engines the exchange of heat is not reversible. That came as startling news to those Newtonian-minded physicists who believed in a determined world. Suddenly, time was no longer reversible: the arrow of time flowed only one way. This put thermodynamics at odds with Newton's universal laws, which claimed that everything was reversible in principle. It was this earth-shattering realization that slowly worked its way into other thinking—even, as we shall see, thinking about layered architecture and how to frame the mind/brain gap problem.

Oddly, by the mid-nineteenth century, atomic theory, the theory that matter is made up of atoms, had been accepted by chemists and put to use, but still wasn't the consensus among physicists. One physicist puzzling over it all was the Austrian Ludwig Boltzmann. He is best known for *kinetic theory*, in which he describes a gas as made up of a large number of atoms or molecules constantly moving, hitting and bouncing off both one another and the walls of their container, producing random chaotic motion. He turned Gassendi's seventeenth-century ideas into the hard science of what is now called statistical mechanics. If the types of molecules and their positions are taken into account, kinetic theory explained the observable macroscopic properties of gases: pressure, temperature, volume, viscosity, and thermal conductivity.

Overall, Boltzmann's huge insight was to further define the disorder of a system (entropy) as the collective result of all the molecular motion. He maintained that with all those atoms bouncing around willy-nilly, the second law was valid only in a *statistical* sense, not in a flat-footed, deterministic sense. That is, whether a particular particle would transfer was unknown. With that parka next to me, my overall system was becoming more disordered. As Michael Corleone said, "It's not personal, it's strictly business."

Boltzmann caused a major ruckus among physicists who still viewed the universe as completely deterministic and ruled by Newton's laws. They firmly believed it was not a statistical universe where mere predictions were as good as it gets. As a consequence, Boltzmann's theory was repeatedly attacked. Sadly, he became so frustrated and depressed by this that he committed suicide in 1906 while on vacation with his family near Trieste, just before his theory was proved unequivocally true.

Physicists to this day are flummoxed by statistical laws. For one thing, Newton's laws are symmetrical with respect to time and are therefore reversible. Clearly, in the determined world defined by Newton, what goes forward can also go backward. Obviously not so with statistical laws. How can something that only happens with a probability, not a certainty, be reversible? It can't, and, on the face of it, these two ways of describing reality are at odds. New thinking was needed to handle this duality. After being slow on the uptake with atomic physics, once they had accepted it, the physicists ran with it and stretched their minds around what this new world showed them. Right off the bat, in 1897, the English physicist Joseph John Thomson discovered and identified the first subatomic particle: the electron. Thomson was both a great physicist and a great teacher. Not only was he knighted and Nobeled for his work, but eight of his research assistants also won their own Noble prizes, as did his son. Included in the group was Niels Bohr, who ultimately presented the idea of complementarity. But I get ahead of myself. Accepting this new world took a bit of convincing.

Max Planck, a German theoretical physicist, was obsessed with the idea of entropy and the second law of thermodynamics. He first believed in its *absolute* validity, not the wishy-washy *statistical* version that Boltzmann was championing. As a champion of Newtonian mechanics, Planck nonetheless realized that entropy presented a problem, since lurking within the concept of increasing entropy was that thorny reality of irreversibility. Planck actually accepted the idea of irreversibility, but he longed to present a rigorous derivation of the entropy law that could justify its irreversibility using classical laws. Like most physicists, he badly wanted a single physical description that explained everything. And old ideas fall hard.

An opportunity presented itself in 1894, when he was commissioned for a special task—to optimize lightbulbs, maximizing the light produced while minimizing the energy used. In order to do this, he had to tackle the problem of what is called *black-body radiation*. We can grasp what this is by going out to the campfire. If you stick a metal shish kebab skewer into the fire, its tip will eventually become red-hot. If it gets even hotter, the color will go from red to yellow to white, then blue. As the interior of the skewer heats up, the surface starts emitting electromagnetic radiation in the form of light, called *thermal radiation*. The hotter the interior (the higher the energy), the shorter the wavelength (and the higher the frequency) of the light that is emitted—thus the color change. Physicists soon posited an idealized object, a "perfect" emitter and absorber that would look black when it is cold, because all light that falls on it would be completely absorbed.

This perfect object is known as a black body, and the electromagnetic radiation it emits is black-body radiation. No one had been able to predict accurately the amount of radiation and at what frequencies such a black body would emit using the classical laws of physics. Newtonian laws worked well when lower frequencies of light (red) were emitted, but as the frequencies increased, the predictions were about as wrong as they could be. After several attempts and failures using purely classical physics, Planck reluctantly turned to the statistical notion of entropy. Once he introduced the idea of "energy elements" and considered that the energy was "a discrete quantity composed of an integral number of finite equal parts,"[12] he was able to come up with an equation that predicted black-body radiation well.

Neither he nor anyone else at the time realized that his radiation law was based on a conceptual novelty, a fundamental change in how to view the world. It was the first foray into the quantum world. Planck's discovery would also suggest that there was neither one fundamental set of laws nor one model of the universe. Some would say it was a nail in the coffin of the Newtonian belief that the world was a determined place.

Interestingly, Planck himself was just tickled at the accuracy of his radiation law and considered that specifying energy quanta was, he said, "a purely formal assumption and I really did not give it much thought."[13] What Planck had stumbled upon, the conceptual novelty that he didn't fully grasp but used as a mathematical trick, is that *microscopic objects behave differently than macroscopic objects*. Whoa! Planck had unwittingly pulled out the brick that held together the foundations of his pet dream: that a single explanation could describe everything. Not only did finding the quantum world change human understanding of the universe, but it was going to become apparent that there are two different layers of reality, and each possesses a different vocabulary and way of doing things. Just as in any complex system, each layer has its own protocol: at the atomic level things worked statistically, but large objects worked just the way Newton said. Lacking the concept of a layered architecture, the epistemologists went nuts. The "How is it we know stuff?" crowd was in disarray, big-time. A single explanation for everything wasn't working. There seemed to be two kinds of explanations for the behavior of matter—a complementarity.

The physicists stumbled on the new idea as they came to understand that light could behave as particles or as waves. There is a complementarity to stuff, a duality. They fought the idea off for decades, but finally they accepted it as truth. Recently, researchers captured an unbelievable picture of a small group of photons as waves and another group behaving as particles at the same time.[14] Although the idea of complementarity is now established in physics, it is not widely seen as a possible foundational idea for thinking about the mind/brain explanatory gap. I think it should be, and first want to look at how physics came to accept its seemingly puzzling reality. Following its acceptance in physics, the idea of complementarity may prove itself to be key to thinking about biology, and about the mind/brain gap in particular.

The Idea of Complementarity

After earning his diploma as a teacher of physics and mathematics, in 1901 the twenty-two-year-old Albert Einstein became a Swiss citizen and struggled to find a job. No educational institution would hire him. He finally snagged a position working in the Bern patent office as a "technical expert, third class," and tutored on the side. In his downtime, he bounced ideas off a couple of pals in a discussion club that they had formed and called the Olympia Academy.

Over the course of 1905, which came to be known as his annus mirabilis, Einstein, now twenty-six, brought physics into a different universe by proposing four huge ideas. He created the quantum theory of light, which stipulates that the energy in a beam of light is, really and truly, in little packets (later called photons), and that energy could be exchanged in only tiny, discrete amounts. The "energy packet" wasn't, after all, some mathematical trick that Planck had ginned up that just produced a good equation. Up until that time it had been debated whether light is a wave or a set of tiny particles. Considering light as a wave explained all sorts of observations, such as light refraction and diffraction, interference, and polarization. It did not, however, explain the photoelectric effect: when light hits a metallic surface, electrons (called photoelectrons in this case) may be jettisoned from the metal's surface.

Initially, physicists didn't see this as any big deal. Assuming the wave theory of light, they figured that the more intense the light (i.e., the higher the amplitude of the wave), the greater would be the energy with which the electrons would be chucked from the metal. But it turns out that this is the opposite of what actually happens. The energies of the emitted electrons are independent of the intensity of the light: bright and dim lights eject electrons from the metal surface with the same energy when the wave frequency is held constant. The unexpected finding was that increasing the wave frequency was what increased the energy with which the electrons were chucked from the surface. This doesn't make sense if light is a wave. It would be like saying that if a big huge ocean wave and a little ripple hit a beach ball, it would fly off with the same energy. Einstein realized that the observed effects could be explained only by light being made up of particles that interacted with the electrons in the metal. In his model, light consisted of individual quanta (which were later called photons) that interracted with the metal's electrons. Each photon carried its own energy. Increasing the intensity of the light increased the number of photons per unit of time, but the amount of energy per photon was the same. Then, a few months later, Einstein was to add to his bonanza year and figured out that light could also be viewed as a wave. Light indeed existed in two realities.

Einstein was unstoppable. He also presented empirical evidence validating the reality of the atom, settling the debate over its existence, and gave the thumbs-up to the use of statistical physics. Putting frosting on the cake, he added the theory of relativity and came up with the famous $E = mc^2$ equation. It took a while for the physics world to cotton to all these ideas, and Einstein didn't immediately gain much recognition. The immediate result of his efforts was solely a promotion in the patent office to "technical expert, class II."

Once physicists got their heads around atomic theory and caught up with the chemists, however, they soon realized that subatomic particles, atoms, and molecules, those submicroscopic universal building blocks of everything, do not follow Newton's laws—they flout them. The smoking gun was that when orbiting electrons lose energy, they don't crash into the nucleus as Newton's laws would predict; they remain in orbit. How could that be?

In 1925 to 1926, quantum theory was further developed by a group of physicists that included Werner Heisenberg, at the University of Göttingen, with frequent trips back and forth to Niels Bohr's institute in Copenhagen, to explain the three big puzzles: the phenomena of black-body radiation, the photoelectric effect, and the stability of orbiting electrons. Physicists, whether they liked it or not (and many, including Planck and Einstein, did not), were jolted out of Newton's deterministic world, the physical "layer" that we inhabit and can see and touch, the one-explanation-fits-all world, to a lower layer, the hidden, nonintuitive, statistical, indeterminate world of quantum mechanics. They were jolted from the world of black-and-white answers to the world of gray answers: a layer with a different protocol that exists simultaneously.

Consider, for example, light reflection. When light photons hit glass, 4 percent are reflected, but the rest are absorbed. What determines which are reflected? After years of research, using multiple techniques, the answer appears to be: chance. It is chance whether a particular photon will be reflected or absorbed. Richard Feynman asked, "Are we therefore reduced to this horror that physics is not reduced to wonderful predictions but to probabilities? Yes we have, that's the situation today. . . . In spite of the fact that philosophers have said, 'It is a necessary requirement for science that setting up an experiment that is exactly similar will produce results exactly the same the second time.' Not at all. One out of 25 it goes up and sometimes it goes down . . . unpredictable, completely by chance . . . that is the way it is."[15] The world of uncertainty. Physicists at the time despised it. Even Einstein, who had opened the door to this uncertain world, wanted to slam it shut. He was having grave doubts about what it implies about a supposedly determined universe and causality, prompting his famous quote "God does not play dice with the universe." Yet, if they were to be good scientists, physicists had to discard their preconceived notions and follow wherever their findings took them.

When considering the wacky quantum world, remember that we inhabit the macro world of Newtonian physics. Common sense, that is, our folk physics based in the macro world, is not going to help us in the quantum world. It is like nothing we have ever experienced. Leave your intuitions at home. They will not be needed and will only be a burden. Feynman entertainingly prepared a physics class for a lecture on quantum behavior with the following disclaimer:

Your experience with things you have seen before is inadequate. Is incomplete. The behavior of things on a very tiny scale is simply different. They do not behave just like particles. They do not behave just like waves. . . . [Electrons] behave as nothing you have seen before. There is one simplification at least, electrons behave exactly the same in this respect as photons. That is, they are both screwy but in exactly the same way. How they behave, therefore, takes a great deal of imagination to appreciate because we are going to describe something which is different from anything that you know about. . . . It is abstract in the sense that it is not close to experience.[16]

He goes on to say that if you want to learn about the character of physical law, it is essential to talk about this particular aspect "because this thing is completely characteristic of all the particles of nature."

The submicroscopic quantum world is hidden from our view. That means in order to learn anything about it, we have to have some type of measurement interaction. This involves some instrument from our macro world, which in turn is made up of atoms, which themselves can react with and disrupt the particles we set out to measure that had innocently been doing their own thing. That disruption is going to set the dynamics of the system off in a direction other than what had been going on before the measurement was made. In short, it was also beginning to look like there was an unavoidable measurement problem. Snooping in on the quantum world was going to be tough and would require some new thinking.

So here we go: it turns out that, just as Einstein discovered, light behaves as both a wave and a particle. A few years later it was also found that the same thing is true about matter: electrons have particle- and wavelike properties, too. Physicists soon accepted the idea that what we perceive in the macro world as continuous (rather than billions of individual atoms), say a dining room table, is merely a simulated averaging process in what the applied mathematician, physicist, and all-around polymath John von Neumann later called "a world which in truth is discontinuous by its very nature." He went on to say, "This simulation is such that man generally perceives the sum of many billions of elementary processes simultaneously, so that the leveling law of large numbers completely obscures the real nature of the individual processes."[17] What "the leveling law of large numbers" means is that the motions of all those particles together cancel one another out, and thus the table stays in one place and isn't shimmying across the floor. When we see a solid table, however, it is an illusion, a symbolic representation, created by our brain, to denote what is really there. It is a very good illusion that delivers good information, which allows us to function effectively in the world.

The Austrian physicist of "cat in a box" fame, Erwin Schrödinger, was also eager to shore up the deterministic world of causality. He developed what came to be known as the Schrödinger equation, a "law" that describes the behavior of a quantum mechanical wave and how it changes dynamically over time. While the "law" is reversible and deterministic, it can't describe the overall system's state. It doesn't take into account the particle nature of an electron, which Schrödinger tried to avoid. The law can't determine where exactly an electron will actually be in its orbit at any given time. Only a prediction, based on probability, can be made about its exact position at any given moment, its so-called quantum state.

In order to know the actual location of the electron, a measurement must be made, and here is where the troubles begin for the die-hard determinists. Once a measurement is made, the quantum state is said to collapse, meaning that all the other possible states the electron could have been in (known as superpositions) have collapsed into one. All the other possibilities have been eliminated. The measurement, of course, was irreversible and had constrained the system by causing the collapse. Over the next couple of years physicists realized that neither the classical concept of "particle" nor that of "wave" could fully describe the behavior of quantum-scale objects at any one point in time. As Feynman quipped, "They don't behave like a wave or like a particle, they behave quantum mechanically."[18]

This is where Niels Bohr, the Danish electron expert and Nobel laureate, came in to help out. After spending a couple of weeks skiing alone in Norway, pondering the dual nature of electrons and photons, he returned having framed the *principle of complementarity*, which this wave-particle duality exemplifies. The principle maintains that quantum objects have complementary properties that cannot both be measured, and thus known, at the same point in time. As Jim Baggott describes in his book *The Quantum Story*, Bohr had

> realized that the position-momentum and energy-time uncertainty
> relations actually manifest the complementarity between the
> classical wave and particle concepts. Wave behavior and particle
> behavior are inherent in all quantum systems exposed to experiment

and by choosing an experiment—choosing the wave mirror or the particle mirror—we introduce an inevitable uncertainty in the properties to be measured. This is not an uncertainty introduced through the "clumsiness" of our measurements, as Heisenberg had argued, but arises because our choice of apparatus forces the quantum system to reveal one kind of behavior over another.[19]

Again, at any one point in time, either the position or the momentum of an electron can be measured and known, but not both—just like its wave-like properties or its particle-like properties. When you make an instantaneous measurement of its position at a single point in time, it actually is in a single location and not moving; thus, its dual nature of also having momentum is compromised. At that point in time, measuring the momentum is not possible. One can only suggest the other measurement with a probability, not a certainty. Complementarity emerges in a system when an attempt to measure one of the paired properties is made. The single system has two simultaneous modes of description, one not reducible to the other.

Bohr worked on this theory for six months and first outlined it in his 1927 Como Lecture, presented to an illustrious group of physicists at a congress commemorative of the one-hundred-year anniversary of the death of Alessandro Volta. Einstein wasn't there and didn't hear it until the next month, when Bohr presented it again in Brussels. Einstein was unhappy with the idea of a dual description and uncertainty. He and Bohr began a lengthy exchange that lasted years. Einstein would come up with a scenario in an attempt to defeat quantum theory, only to have Bohr present an argument consistent with quantum theory that overcame it. Since then, many proposals have been made and experiments done attempting to bolster Einstein's side of the debate;[20] all have been unsuccessful. While unpopular among those with a determinist bias, a version of Bohr's complementarity remains undefeated.

At the root of their argument is what objectivity means and what physics is about. Robert Rosen has explained just what is at stake:

Physics strives, at least, to restrict itself to "objectivities." It thus presumes a rigid separation between what is objective, and thus falls directly within its precincts, and what is not. Its opinion about whatever is outside these precincts is divided. Some believe that whatever is outside is so because of removable and impermanent technical issues of formulation; i.e., whatever is outside can be "reduced" to what is already inside. Others believe the separation is absolute and irrevocable.[21]

Bohr belonged to the latter group and argued that whether we see light as a particle or a wave is not inherent in light but depends on how we measure and observe it. Both the light and the measurement apparatus are part of the system. For Bohr, the classical world is just too small to describe all material reality. For him, the universe and all it contains is way more complex and requires more than a single layer with a protocol made up of the laws of classical physics. Rosen notes that Bohr changed the concept of "objectivity" itself, from pertaining only to what is inherent entirely in a material system to what is inherent in a system-observer pair. Einstein couldn't swallow this and threw his lot in with classical physics, which ignores the measurement procedure and looks at its outcome as inherent in light. For Einstein, something is objective only if it is independent of how it is measured or observed. Rosen concludes, "Einstein believed that there was such knowledge, immanent alone in a thing, and independent of how that knowledge was elicited. Bohr regarded that view as 'classical,' incompatible with quantum views of reality, which always required specification of a context, and always containing unfractionable information about that context."[22]

Bohr's principle of complementarity is not just some interesting bit of scientific bickering with Einstein. We are going to see that it is fundamental to understanding the mind/brain gap.

8.

NON-LIVING TO LIVING
AND NEURONS TO MIND

In the beginning of heaven and earth there were no
symbols. Symbols came out of the womb of matter.

—Lao-tzu

An important stage of human thought will have been
reached when the physiological and the psychological,
the objective and the subjective, are actually united.

—Ivan Pavlov

N THE SEARCH to understand matter more precisely, physicists stumbled
upon complementarity, the concept that all matter can exist in two different
states at the same time. Accepting this duality did more than simply push the
boundaries of physics. It demanded new thinking, beyond what we could
imagine from our own experience of natural phenomena, to understand the
natural world. Today, new thinking and imagination stretching are needed
by those studying mind/brain duality. We need someone who will think outside of the twenty-five-hundred-year-old box stuffed full of human intuitions
and conventional wisdoms. We need someone who has struggled with the
ebb and flow of modern physics, who recognizes the importance of comple-

mentarity. We need someone who thinks the last few millennia have been frittered away by philosophers looking for answers in the wrong place: the highly evolved human brain. We need someone like Howard Pattee, a Stanford-educated physicist who waded into theoretical biology during his stunning career at SUNY Binghamton. Pattee, who has been a keen observer of human thought, feels that philosophers have approached the mind/brain divide from the wrong end of evolution.[1] Over the course of his life, Pattee has come to a startling conclusion: duality is a necessary and inherent property of any entity capable of evolving.

Howard Pattee doesn't concern himself with the gap between the material brain and the immaterial mind. He delves much deeper. The original gap, the true source of the problem, was there long before the brain. The mother of all gaps is that which lies between lifeless and living matter. The fundamental problem started at the origin of life on Earth. We should not focus exclusively on the gap between the physical brain and the ethereal mind. We must understand the difference between the conglomerations of matter that produce a material object that is lifeless and the conglomerations of matter that produce a thing full of life. The gap between living and nonliving is at the root of the gap between the mind and the brain and offers a framework for addressing the mind/brain problem.

Pattee's idea takes a little time to get used to and comes with an important warning: if we want to understand the idea of consciousness, something fully formed in evolved living systems, we must first understand what makes a living system alive and evolvable in the first place. What happened that split things up into two domains, one living, one not? Most of us have thought about the "life springing from matter" problem for a few seconds, then dismissed it as too difficult and gone about our business poking and measuring what is in front of us. Not Pattee. He was captivated by questions about the beginnings of life when he was an adolescent, first dipping his toes into scientific waters at boarding school in the late 1930s. His science teacher and headmaster, Dr. Paul Luther Karl Gross, gave him a book for summer reading. It was not the usual light fare for summer. It was *The Grammar of*

Science (first published in 1892) by Karl Pearson, a brilliant mathematician and statistician.

At the time, Pattee wondered why he was being given a seemingly out-of-date science book written before the days of quantum theory. However, in the chapter "The Relation of Biology to Physics," Pattee found a question that would motivate his thinking for decades: "How . . . is it possible for us to distinguish the living from the lifeless if we can describe both conceptually by the motion of inorganic corpuscles?"[2] Pattee saw the logic of the question, but he also understood that evoking the same laws to describe both animate and inanimate matter was not a good enough explanation. In fact, it was no explanation at all. There had to be more to the story.

The Headache of Quantum Mechanics

Pattee was very lucky to have such an exceptional headmaster. Dr. Gross stimulated his students' thinking by taking them to cutting-edge scientific events, including the Nobel laureate Linus Pauling's evening lectures at Caltech. One evening, Pattee heard Pauling describe Schrödinger's famous cat paradox, the notion that a cat could be both dead and alive at the same time. It went like this: A cat is penned up in a steel chamber, along with a very tiny bit of radioactive substance and a Geiger counter to measure the radiation emanating from the substance. Due to the radioactive decay rate of the substance, there is a fifty-fifty chance that, within an hour, none of the atoms will decay. However, there is, of course, a fifty-fifty chance that an atom will decay, and if it does the Geiger counter tube will discharge. In this contraption, the Geiger counter discharge will release a hammer that shatters a small flask of hydrocyanic acid, killing the cat. This elaborate setup yields a scenario in which there is a 50 percent chance that, by the end of the hour, the cat is alive, and a 50 percent chance that the cat will be dead. Odd, but fair enough. However, in quantum mechanics, this phenomenon would not be expressed as the probabilities of two outcomes. It would be expressed as what is called a *psi function*, a description of the entire quantum state of the system. The psi function of Schrödinger's poor cat would have the living and the dead cat smeared out in equal parts! Pattee sat there puzzled. How could quantum mechanics, a theory that so powerfully explained all of chemistry and most of physics, produce such complete nonsense as the Schrödinger cat problem? It launched him into a lifelong journey seeking resolution to this dilemma.

The puzzle that perplexed the young Pattee is referred to as the *measurement problem*. We discussed in chapter 7 that a quantum system has paired complementary properties that cannot be measured simultaneously. Measurements present additional challenges at the quantum level for three reasons. First, measurement requires an observer, a subject or agent who is separate from the object being measured. Second, the measurement process (being irreversible) is not governed by the classical laws of physics. Third, measurement has arbitrary aspects: the observer chooses when, where, and what to measure, as well as the symbols (themselves arbitrary) used to express the measurement. Measurement is a selective process in which most aspects of the thing being measured are actually ignored. Let's say I want to describe you. What one measurement should I choose to capture you? I decide to measure your mass. I will take one weight measurement and use this to describe you over your lifetime. When should I take it? When you are an infant? An adult of twenty, thirty-five, sixty? The day before or after Thanksgiving? Which is most representative? Is weight alone a good measure of you? What about height and weight together? The measurement itself may be precise and objective, but the *process* of measurement is subjective.

The process of measuring is arbitrary, which means it can't be described by objective laws, whether they be quantum or classical. This presents a problem for all of physics, not just the quantum variety. To make predictions about the future state of a system, a physicist must know the initial conditions of a system. How? Through measurement of the initial conditions of a system. Yet this measurement is arbitrary and, in making it, the physicist interferes with the initial conditions. The subjectivity of the initial measurement is commonly ignored by determinists when they assume that the world is completely predictable. But there is no escaping it. No matter how hard you try to be an objective observer, by the mere fact of measuring you are introducing subjectivity into the system. The "measurement problem" deals a huge blow to physics, but it might be just what neuroscience needs.

The *Schnitt* and the Origins of Life

Physicists refer to the inescapable separation of a subject (the measurer) from an object (the measured) as *die Schnitt*. (What a great word!) Pattee calls "this unavoidable conceptual separation of the knower and the known, or the symbolic record of an event and the event itself, the *epistemic cut*."[3] There is a world of actions that exists on the side of the observer with the observer's record of an event. There is also a separate world of actions on the side of the event itself. This sounds confusing, but think of the explanatory gap between your subjective experience of an event (I had so much fun body-surfing) and the event itself (A person went swimming in the ocean). Alternately, you can think of the explanatory gap between the same subjective experience (This is fun) and the goings-on within the brain (Some neurons fired while a person was swimming in the ocean). These are all just versions of the subject/object complementarity seen in physics. Here is the really wild part: Who's measuring the events? To examine the difference between a person's subjective experience and objective reality, do we need a scientist? Who's measuring the scientist?

Pattee points out that neither classical nor quantum theory formally defines the subject, that is, the agent or observer that determines what is measured. Physics, therefore, does not say *where* to make the epistemic cut.[4] Quantum measurement does not need a physicist-observer, however. Pattee argues that other things can perform quantum measurements. For example, enzymes (such as DNA polymerases) can act as measurement agents, performing quantum measurement during a cell's replication process.[5] No human observer is needed.

The nonsense of Schrödinger's cat, which even the adolescent Pattee wanted no part of, arose not from the cat or the Geiger counter, but from the human performing the bizarre experiment. In Schrödinger's thought experiment, the cat is described as a psi function, dead and alive at the same time. This situation persists until we open the box and make a measurement; that is, we observe whether the cat is alive or dead, but no longer both. The result of the measurement intervention (the human opening the box) appears to be instantaneous and irreversible, and the physical representation of the result (live cat or dead cat) is arbitrary. Yet how can this be when at the same time, all microscopic events are assumed to obey reversible quantum dynamical laws (e.g., Schrödinger's equation)? Pattee notes that it was this inadequate model of measurement that prevented the state of Schrödinger's cat from being known before it was observed. He states, "It was the belief that human consciousness ultimately collapsed the wave function that produced the problem of Schrödinger's cat."[6] In fact, Schrödinger wrote his cat scenario specifically to illustrate that this notion was ridiculous. He hoped to illustrate that quantum superposition could not work with large objects, such as cats (or dogs or you, for that matter).

For Schrödinger, the joke was on us. He was trying to point out that there is something missing in our understanding. Pattee got it (in high school) and buckled down to attack the problem. Where should we put the cut, the gap, *die Schnitt*? With his consuming interest in the origins of life, he came to realize that human consciousness was way too high a layer in the architecture of all living organisms to put the epistemic cut between the observer and the observed, between the subjective experience and the event itself. There are umpteen layers between subatomic particles and human brains. There are plenty of layers between subatomic particles and brains in general (cat or mouse or fly or worm). Putting the major epistemic cut that high led to the nonsense of Schrödinger's cat existing as a quantum system. There was no pussyfooting around for Pattee: "I have taken the point of view that the question of what constitutes an observation in quantum mechanics must arise long before we reach the complexity of the brain. In fact, I propose . . . that the gap between quantum and classical behavior is inherent in the distinction between inanimate and living matter."[7]

There you have it. Pattee proposes that the gap resulted from a process equivalent to quantum measurement that began with self-replication at the origin of life with the cell as the simplest agent.[8] The epistemic cut, the subject/object cut, the mind/matter cut, all are rooted to that original cut at the origin of life. The gap between subjective feeling and objective neural firings didn't come about with the appearance of brains. It was already there when the first cell started living. Two complementary modes of behavior, two levels of description are inherent in life itself, were present at the origin of life, have been conserved by evolution, and continue to be necessary for differentiating subjective experience from the event itself. That is a mind-boggling idea.

A Life in Symbols: Von Neumann Shows the Way

Living matter seems to be playing an entirely different game than non-living matter, even though they are both made from the same stuff. Why is living matter different from non-living matter? Is it simply cheating, somehow violating the physical laws that we've come to understand govern non-living matter? Pattee argues that living matter is distinguished from non-living matter by its ability to replicate and to evolve over the course of time. So what does it take to replicate and evolve?

John von Neumann was a Hungarian-born mathematical genius and an electrifying bon vivant whose intellectual contributions were as vast as his appetite for life. Born into Jewish aristocracy in Budapest, he died receiving last rites from a Catholic priest—he quipped he had adopted Pascal's wager!*

---

*Pascal was interested in probability theory. The gist of the wager concerns what degree of risk is acceptable if the consequence of being wrong is dire. Is it worth it to risk the substantial consequences of believing God does not exist, if indeed God does?

In the intervening years, he moved to Princeton's Institute for Advanced Studies, where he reportedly drove Einstein crazy playing German marching music on his gramophone at full volume.

The intellectual atmosphere of the times was vibrant. Schrödinger had delivered his history-making lecture on "What is life?" in Dublin in 1943, in which he made the suggestion that a "code script" was somehow captured in the molecular mechanisms of a cell. By the late 1940s, von Neumann had assigned himself the question of life as a thought experiment as well. What is life? Well, what do living things do? One answer is, they reproduce. Life makes more life. However, logic told him that "what goes on is actually one degree better than self-reproduction, for organisms appear to have gotten more elaborate in the course of time."[9]

Life did not *just* make more life. Life could increase in complexity; it could evolve. Von Neumann became increasingly interested in what an evolvable, autonomous, self-replicating machine ("an automaton") would logically require when placed in an environment with which it could interact. His string of logic led him to the conclusion that the automaton needed a description of how to copy itself and a description of how to copy that description so it could hand it off to the next, freshly minted automaton. The original automaton also needed a mechanism to do the actual construction and copy job. It needed information and construction. However, this would cover only replication. Von Neumann reasoned that he had to add something in order for the automaton to be able to evolve, to increase in complexity. He concluded that it needed a symbolic self-description, a genotype, a physical structure independent of the structure it was describing, the phenotype. Linking the symbolic description with what it refers to would require a code, and now his automatons would be able to evolve. We will see why in a bit.

It turned out that von Neumann was right on the money. He correctly predicted how cells actually replicate before Watson and Crick did. From the get-go, at the origin of life, at the single-molecule level, when DNA was still a twinkle in Mother Nature's eye, evolvable self-replication depended on two things: (1) the writing and reading of hereditary records, which were in some type of symbolic form, and (2) the sharp distinction between the description and the construction processes. After this little thought experiment, von Neumann was off to other endeavors and puzzles. However, von Neumann left his job half done: He did not address the physical requirements for implementing his logic. Rubbing his hands together, Pattee took up the challenge.

The Physics of Symbols: Pattee Presses Onward

We tend to think of symbols as being abstract, not something that is subject to physics. However, as scientists, we are physical beings looking for physical evidence that abides by physical rules and laws. There must be a physical manifestation of von Neumann's symbols. What Pattee calls the *physics of symbols* produces certain problems. The first problem lies in the writing and reading of hereditary records, the informational description. A description involves a recording process, and, as we learned in the preceding section, a record is an irreversible measurement that requires some measurer. Pattee realized that the informational description at the origin of life comes face-to-face with the measurement problem in quantum mechanics. Measurements are subjective, meaning they can't be described by objective laws, whether they be quantum or classical. Any living thing that "records" information is introducing a form of subjectivity into the system.

The second problem is the relationship between the genotype and the phenotype. For example, when we consider DNA, the *genotype* is the DNA sequence that contains instructions for the living organism. The *phenotype* is the observable characteristics of an organism, such as its anatomy, biochemistry, physiology, and behavior. The genotype interacts with the environment to produce the phenotype. To put this in an everyday situation, consider the blueprint as a house's genotype and the actual house its phenotype. The phenotypic construction process is the building of the house using the blueprint as information about what and how to do it. The phenotype is related to the genotype that describes it, but there is a world of physical difference between the genotype and the phenotype and even the phenotypic construction process. For one, the genotype is non-dynamic; it is a quiescent, one-dimensional sequence of symbols (DNA's symbols are nucleotides) that has no energy or time constraints. Like a blueprint, it can sit around for years, as you have probably learned from watching *CSI*. The genotype dictates what should be constructed (perhaps a really cute dog), but the DNA itself does not look or act anything like a cute dog. On the other hand, the phenotype (the cute dog) is dynamic and uses energy, especially if it is a border collie.

The phenotypic construction process is also related to the genotype. Just as a blueprint constrains the builder from adding turrets to a house, the genotype constrains how many tails that cute dog is going to have. How do these relate? What is the relationship between a blueprint, a house, and the intervening pouring of concrete and pounding of nails? The hereditary records contain not only information that specifies *what* to build but also information that specifies *how* to build it. Somehow, the information about what and how has been "recorded" in some type of symbolic form. There is a gap between the subjectively recorded symbol (genotype) and the phenotypic construction process and the phenotype. The symbols have to be translated into their meaning for construction to begin. If we think of a layered architecture, this would be the protocol between two layers. Pattee proposes that it was from the control interface between these two layers that the epistemic cut arose. In the case of DNA, the bridge between the genotype and phenotype is the genetic *code*. In the case of the blueprint, the bridge is the building contractor translating the blueprint for the construction workers.

Pattee extended von Neumann's logic by contending that the symbols themselves, which make up the instructions (the hereditary record), *must have a material structure* and that the material structure, during the phenotypic construction process (the building of the new automaton), constrains the process in a way that follows Newton's laws. There are no magic tricks here. The symbols are actual physical structures, a chain of nucleotides that obey the classical laws of physics.

Here's the kicker: a symbol, whether it is a sequence of nucleotides in DNA, a sequence of Morse code, or a sequence of mental simulations, is arbitrary. The arbitrariness of symbols can be easily grasped when looking at the ever-changing world of slang. For example, "Benjamins," "simoleons," and "dough" have all been popular, though arbitrary, symbols for money. And each language has its own set of symbols, as the comedian Steve Martin once warned anyone traveling to Paris, "*Chapeau* means hat, *oeuf* means egg. It's like those French have a different word for everything."[10] The problem is, Newton doesn't do arbitrary. If Newton's inflexible laws ruled symbols, then every person, all of the world, would use the exact same word to represent the concept of money, every time, for all of eternity. Sadly for Newton, there are numerous possible symbols that can be selected to convey information. Each would have different properties, different pros and cons, but since the symbols are distinct from the thing itself, there is not a one-to-one mapping.

You may object that DNA is arbitrary in the same way that language is, arguing that there are physicochemical constraints. But the selection of a symbol is not governed by physical *laws* but by a *rule*: select the symbol that carries the most useful and reliable (stable) information for the system. We will see in a bit that the components of DNA itself were selected from a range of competitors for doing a better job constraining the function of the system it belongs to. And if it is stable it can be transmissible. The current components of DNA are what Pattee refers to as frozen accidents. The current symbol embodies the history of its successful versions over multiple timescales (time independent), not that of its current action. So, back to money: if only a couple of people refer to money as "Bettys," it is not a reliable symbol and won't be selected or transmitted.

This is a bit confusing because in our social world, we often use "rules" and "laws" interchangeably. We call such things as the rules for driving "laws." Pattee explains there is a basic and extremely important distinction between laws and rules in nature.[11] Laws are inexorable, meaning they are unchangeable, inescapable, and inevitable. We can never alter or evade laws of nature. The laws of nature dictate that a car will stay in motion either until an equal and opposite force stops it or it runs out of energy. That is not something we can change. Laws are incorporeal, meaning they do not need embodiments or structures to execute them: there is not a physics policeman enforcing the car's halt when it runs out of energy. Laws are also universal: they hold at all times in all places. The laws of motion apply whether you are in Scotland or in Spain.

On the other hand, rules are arbitrary and can be changed. In the British Isles, the driving rule is to drive on the left side of the road. Continental Europe's driving rule is to drive on the right side of the road. Rules are dependent on some sort of structure or constraint to execute them. In this case that structure is a police force that fines those who break the rules by driving on the wrong side. Rules are local, meaning that they can exist only when and where there are physical structures to enforce them. If you live out in the middle of the Australian outback, you are in charge. Drive on either side. There is no structure in place to restrain you! Rules are local and changeable and breakable. A rule-governed symbol is selected from a range of competitors for doing a better job constraining the function of the system it belongs to, leading to the production of a more successful phenotype. Selection is flexible; Newton's laws are not. In their informational role, symbols aren't dependent on the physical laws that govern energy, time, and rates of change. They follow none of Newton's laws. They are lawless rule-followers! What this is telling us is that symbols are not locked to their meanings.

Symbols lead a double life, with two different complementary modes of description depending on the job they are doing. In one life, symbols are made of physical material (DNA is made of hydrogen, oxygen, carbon, nitrogen, and phosphate molecules) that follows Newton's laws and constrains the building process by its physical structure. However, in the other life, as repositories of information, the symbols ignore these laws. The double life of symbols has largely been ignored. Those interested in information processing ignore the objective material side, the physical manifestation of the symbol. Molecular biologists and determinists, interested in only the material side, ignore the subjective symbolic side. By claiming just one aspect, neither studies their full, complementary character. Which is not only a shame but a scientific travesty because, as we discussed earlier, for a self-reproducing and evolvable form of life to exist, physical symbols must perform both roles. Pattee argues that either one alone is insufficient. To avoid either side of the link results in missing the link altogether. He boldly states, "It is precisely this natural symbol-matter articulation that makes life distinct from non-living physical systems."[12]

The Genetic Code Is a Real Code

To better understand this symbol-matter articulation and what its implications are for our quest, let's look closely at DNA, which best exemplifies the symbol-matter structure in a living system. First, however, we need a quick primer in biosemiotics to understand about symbols in living systems. Our guide is Marcello Barbieri, a theoretical biologist from the University of Ferrara.

Semiotics is the study of signs (a.k.a. symbols) and their meanings. Basic to the field is that, by definition, a sign is always linked to a meaning. As we already grasped from Steve Martin and his problems in Paris, there is no deterministic relationship between a sign and its meaning. An egg is an egg, whether it resides in the United States or in France, but we can call it different things. The object is distinct from its symbolic representation (the sound "egg" or "*oeuf*") and our understanding of the symbol. Barbieri notes that the relationship between a sign and its meaning is established by a code, a conventional set of rules that establish the correspondence between signs and their meanings. The code is produced by some agent, the codemaker. A semiotic system originates with the codemaker making the code. Thus, Barbieri notes, "a semiotic system is a triad of signs, meanings, and code that are all produced by the same agent, i.e., by the same codemaker."[13]

Biosemiotics is the study of signs and codes in living systems. Foundational to the field is the notion that "the existence of the genetic code implies that every cell is a semiotic system."[14] Barbieri states that modern biology has not accepted this fundamental premise of biosemiotics, because there are three concepts at the heart of modern biology that are not compatible with it. First is the description of the cell as a computer. In this metaphor, genes (biological information) are seen as the software and the proteins the hardware. Computers have codes, but they are not semiotic systems, because the codes come from outside the system via a codemaker, and, as we learned above, a semiotic system *includes* the codemaker. The cell-as-computer concept also contends that the genetic code arose from a codemaker outside the system—natural selection. Under this description, living things are not semiotic systems and "genetic code" is simply a metaphor.

The second concept at issue between modern biology and biosemiotics is physicalism, the notion that everything is reducible to physical quantities. Biologists demand that things (DNA, molecules, cells, organisms) abide by laws that determine their behavior. A semiotic code has the nondeterminist, wishy-washy aspect of rules, not deterministic physical laws linking the symbols inexorably to their meaning. The third source of discord is the conviction (or lack thereof) that all biological innovation is the result of natural selection.

Barbieri argues that biologists are overlooking something basic when making those fundamental assumptions: they are ignoring the origin of life. Evolution by natural selection requires the copying of genetic records and the construction of proteins, but these processes themselves had to originate somehow. Barbieri points out that genes and proteins in living systems are fundamentally different from all other molecules, primarily because they are produced in a totally different way.

The structure of molecules in the inorganic world, the world of objects such as computers and rocks, is determined by the bonds that form spontaneously between their atoms. The bonds themselves are determined by internal factors, the chemical and physical characteristics inherent to the atoms. Happily deterministic.

Not so in living systems. Genes are elaborate strings of nucleotides, and proteins are elaborate strings of amino acids. These strings do not come together spontaneously in a cell. It is not love at first sight drawing them together by an irresistible chemistry. Instead, they are cobbled together by the actions of an entire class of molecules, a whole system of ribonucleic acid (RNA) and protein matchmakers that help them. Barbieri points out that this is highly significant for its implications concerning the origin of life.

Primitive "bondmaker" molecules, early precursors of the RNA system, which bind nucleotides together, came into existence way before the first cells. So, too, did bondmaker molecules that developed the ability to join nucleotides together following a template—"copymakers." These bondmakers and copymakers came into existence through random molecular resorting. It was the existence of copymaker molecules that set the process of evolution into motion. Natural selection chiseled living things into existence, but the requisite molecules for evolution—the bondmakers and copymakers—existed before life itself.

Barbieri chides that "natural selection is the long-term result of molecular copying and would be the sole mechanism of evolution if copying were the sole basic mechanism of life."[15] But it isn't. While genes can be their own template and copy themselves, proteins cannot. Proteins cannot be made by copying other proteins. The tricky thing is that only molecules that can copy can be inherited, so the information about how to make the proteins had to come from the genes. Barbieri notes that the outstanding feature of the very early protein makers "was the ability to ensure a specific correspondence between genes and proteins, because without it there would be no biological specificity, and without specificity there would be no heredity and no reproduction. Life, as we know it, simply would not exist without a specific correspondence between genes and proteins."[16] The specific correspondence he is talking about is a code. The code had to be there first, before natural selection was set in motion.

Here's the interesting thing for us: if that correspondence were not a code, but determined by stereochemistry, which is what was first assumed, it would be automatic—and thus deterministic. But that is not the mechanism, which was a surprise to biologists. The bridge between the genes and the amino-acid sequences they code that make up proteins is provided by transfer RNA molecules. These molecules have two separate recognition sites: one is for a codon (a group of three nucleotides), and another for an amino acid, thus binding the two. This, too, could be the setup for an automatic correspondence between a codon and a specific amino acid if one recognition site physically determined what bound to the other, but it doesn't. The two sites are independent of each other and physically separated. Barbieri notes, "There simply is no necessary link between codons and amino acids, and a specific correspondence between them can only be the result of conventional rules. Only a real code, in short, could guarantee biological specificity, and this means that in no way the genetic code can be dismissed as a linguistic metaphor." So he leaves us with this conclusion: "The cell is a true semiotic system because it contains all the essential features of such systems, i.e., signs, meanings and code all produced by the same codemaker."[17]

Evidence for such a biosemiotic system that contradicts modern biology's foundational concepts is hot off the press. Recently scientists have found that cephalopods (the family that contains the octopus) can recode their RNA. RNA molecules have the privilege of establishing codes with DNA (in the part of the RNA that recognizes the three-nucleotide DNA codon sequence) and also with proteins (in the separate part of the RNA that recognizes the amino acid). Recoding the RNA means that new proteins can be constructed while the DNA sequence of symbols stays the same. The collective result is the destruction of the one-to-one gene-to-protein correspondence. Recoding allows a single octopus gene to produce many different types of proteins from the same DNA sequence.[18] This is a big deal. It is evidence against the three concepts in biology that dismiss semiotic systems in living organisms. The system can change its code. The system has an internal codemaker that can produce biological innovations—new proteins—but not via natural selection. It illustrates the arbitrariness of the connection of a symbol with its meaning in a living system.

If symbols within living systems are arbitrary and RNA is the codemaker, why the preoccupation with DNA? Why has DNA had a monopoly on molecular symbolism over the past few hundreds of millions of years? In its physical manifestation, DNA is extremely structurally stable, unlike RNA. This has helped DNA remain the symbolic structure of choice throughout evolution. However, while the DNA in our cells and the cells of other living organisms is now very stable, the structure of DNA did not start out that way at the very origins of life. Random shuffling and re-sorting of molecules, through the *irreversible* and *probabilistic* process of natural selection, generated molecules resembling nucleotide bases. Through subsequent shuffling, successful DNA components and sequences survived and replicated.

However, what do we mean by "successful" when we talk about DNA? DNA is made up of four different nucleotides. Genes are strings of particular nucleotide combinations that act as the symbolic description, the recipe, for making proteins. What would make a DNA sequence successful? Do we mean successful in remaining physically stable over the lifetime of an organism? Do we instead mean successful in reliably encoding information for the successful replication of the organism? We mean both. DNA, the hereditary memory structure when constraining the construction of DNA, abides by Newton's laws and remains thermodynamically stable in the aqueous environment of the cell based on the properties of its nucleotide bases. However, in its informational (subjective) mode, DNA follows rules, not the laws of physics. The sequences of bases that survive have been selected by evolution according to a rule: pick the most reliable and useful information for the organism's survival and reproduction. The nucleotides that make up DNA and carry information in symbolic form were selected and, although they are arbitrary, have been conserved in a stable form through the process of evolution for doing and continuing to do a good job, unlike professors who can coast after getting tenure.

During replication, those nucleotides are read and translated into linear strings of amino acids (which make up enzymes and proteins) by a rule-governed process. The set of rules is called the genetic code. The DNA contains the sequence, but the code is implemented by RNA molecules. Certain

DNA sequences, called codons, which are made up of three nucleotides, symbolize certain amino acid sequences. There is no ambiguity, but there is also not just one codon for each amino acid. For example, six different codons symbolize arginine, but only one codon symbolizes tryptophan. But the components of the DNA sequence (the symbol) do not resemble the components of the amino acid sequence (its meaning), just as the words that symbolize the components of a recipe do not resemble the components themselves.

When a sequence of DNA has been translated into a chain of amino acids, it is the (temporary) end of DNA's instructional activity. It is not the end, however, of the constraints this chain of symbols has placed on the material structure of those amino acids. After the amino acid chain is constructed (remember the amino acids are not bonding to one another spontaneously), it folds itself, forming weak chemical bonds between the molecules that act almost like weak magnets. What bonds are formed and how the chain folds is dependent on which amino acid is where, all dictated by the symbolic description. This is the tricky part. Once those amino acids are placed, the bonds that form are determined by physical laws. Certain amino acids hide from water, while others love it; certain amino acids stick to one another, sometimes quite ardently. The interaction of the amino acid chain with its environment folds the chain into a three-dimensional structure, a protein.[19] Folding is what transforms the rule-following linear, one-dimensional sequence of amino acids into a law-abiding three-dimensional, dynamic, and functional control structure (a protein).

Proteins of course obey the causal laws of physics and chemistry. Yet arbitrary, symbolic information in the DNA sequences is what determines both the material composition and the biochemical function of proteins. Pretty impressive. DNA is the primeval example of symbolic information (the nucleotide sequences) controlling material function (the action of enzymes), linked by a rule-governed code, just as von Neumann had predicted must exist for evolving, self-reproducing automatons. But wait: What made the protein? DNA had the information that was decoded to make the protein, but what kicked off the process? Answer: another protein. The DNA strands had to be ripped apart by an enzyme (a protein) to get the whole replication process going in the first place. It was a newly minted enzyme that pried those DNA strands apart. It's the old chicken-and-egg problem: without catalytic enzymes to break apart the DNA strands, DNA is simply an inert message that cannot be replicated, transcribed, or translated, but without DNA there would be no catalytic enzymes. Bohr's complementarity—two complementary parts, two modes of description, making up a single system.

Von Neumann, in his thought experiment about self-replication, had written that he had avoided the "most intriguing, exciting, and important question of why the molecules or aggregates that in nature really occur . . . are the sorts of thing they are, why they are essentially very large molecules in some cases but large aggregations in other cases."[20] Pattee suggested that it is the very size of the molecules that ties the quantum and classical worlds together: "Enzymes are small enough to take advantage of quantum coherence to attain the enormous catalytic power on which life depends, but large enough to attain high specificity and arbitrariness in producing effectively decoherent products that can *function* as classical structures."[21] Quantum coherence basically means that subatomic particles sync together to "cooperate" to produce decoherent products, which are particles that do not have quantum properties. Pattee notes that there is now research that supports his proposal that enzymes require quantum effects[22] and that life would be impossible in a strictly quantum world.[23] Both are needed: a quantum layer and a classical physical layer.

The Snake Eating Its Tail: Semiotic Closure

Von Neumann made it clear that his automaton would need to replicate. In order to self-replicate, the boundaries of the self must be specified. To make a "self," you need parts that implement description, translation, and construction. To make another self, you need to describe, translate, and construct the parts that describe, translate, and construct. This self-referential loop is not just a headache. It amounts to a logical closure that, in fact, defines a "self."

Pattee calls the physical conditions that are required for this exceptional interdependence of symbol-matter-function *semiotic closure*. He emphasizes that to physically execute this closure, the symbolic instructions *must have a material structure*. There can be no ghost in the system, and the physical structure must constrain all the lawful dynamic processes of construction following Newton's laws. The closing of the semiotic loop, the physical bonding of the molecules, is what defines the limits of the "self," the subject, in "self-replication." No random structures floating around are being incorporated; the limits have been set. This does not imply that the cell is somehow self-aware. However, there can be no self-awareness without a self. The first steps must be toward a delimited self. The subsequent destinations of self-awareness, self-control, self-experience, self-consciousness, and self-absorption are all farther down the road.

Semiotic closure must be present in all cells that self-replicate. Sure, "the self" became more elaborate through evolutionary processes, but even a cell follows Dirty Harry's advice and "knows its limitations." Whatever complex physical processes close the symbol-matter loop, they are the bridge that spans the physicist's *Schnitt*, the explanatory gap, the chasm between the subject and object. They are the protocol between the quantum layer and the Newtonian layer. The processes that close the symbol-matter loop unite the two modes of description, spanning the gap that originated at the origin of life. The implication is that the gap between subjective conscious experience and the objective neural firings of our physical brains, those two modes of description, may be bridged by a similar set of processes, and it could even be possible that they are occurring inside cells.

The Surrender and the Truce

In the early days of quantum physics, Niels Bohr presented the principle of complementarity as a white flag, an attempt to explain the dual nature of light (wave-particle duality). The complementarity principle accepts both objective causal laws and subjective measurement rules as fundamental to the explanation of the phenomena. Bohr emphasized that while two modes of description were necessary, this did not correspond to a duality of the system under observation. The system itself was unified. It was both at the same time. Two sides of one coin.

This is what makes it tricky for us to understand, if we even do. Indeed, Richard Feynman said, "I think I can safely say that nobody understands quantum mechanics." Bohr made an analogy with the distinction between subject and object that extended all the way to mind and matter in his 1927 Como Lecture: "I hope . . . that the idea of complementarity is suited to characterize the situation, which bears a deep-going analogy to the general difficulty in the formation of human ideas, inherent in the distinction between subject and object."[24]

Pattee, however, is bolder. He sees more than an analogy. He sees complementarity as an epistemological necessity that began at the origin of life and extends to all evolved levels. Its essence is not merely the recognition of the subject/object split, but "the apparently paradoxical articulation of the two modes of knowing."[25] That paradox has had philosophers and scientists in a dualist hullabaloo for more than a couple of thousand years. If they keep going at it as they are currently, they'll fight for a couple of thousand more. The two modes of investigation, the two phenomena they unearthed, are not described by the same set of physical laws. Pattee chuckles that the objective mode has led "reductionists to claim that life is nothing but ordinary physics, which indeed it is as long as one is not willing to consider the subjective problems of measurements and descriptions. . . . What the principle of complementarity says is that using only this one objective mode of description not even physics is reducible to this mode!"[26]

Just as scientists and philosophers had to accept the fact that the world wasn't flat, we are going to have to deal with the principle of complementarity as it applies to the mind and brain. The principle of complementarity is still controversial because it butts heads with the belief that the best explanation of something is a single explanation. Yet the single-explanation fallacy fizzled a hundred years ago in physics with the discovery of the quantum world. The micro world follows different laws than the macro world. They inhabit different layers of description, and one is not reducible to the other.

Those who subscribe to the single-explanation gold standard simply are ignoring the realities of physics. Pattee laments that complementarity's "acceptance in quantum mechanics only came about because of the failure of every other interpretation."[27] This echoes Sherlock Holmes's famous saying "When you have eliminated the impossible, whatever remains, no matter how improbable, must be the truth." Pattee wonders if complementarity's acceptance into biological and social theories awaits the same agonizing fate. As Richard Feynman once quipped, "You don't like it then go somewhere else. . . . Go to another universe where the rules are simpler, philosophy more pleasing, more psychologically easy."[28] Just because you don't like the idea doesn't mean it isn't the way things are.

Summing Up

Living matter is distinct from inanimate matter because it has taken an entirely different course. Inanimate matter abides by physical laws. Life from the get-go has thrown its lot in with rules, codes, and the arbitrariness of symbolic information. The distinction between, and the interdependence of, symbolic information and matter has made open-ended evolution possible, resulting in life as we know it. Information of past successful events was cached in records made up of symbols. These records are themselves measurements that are inherently probabilistic in nature. Nonetheless, life is dependent on these arbitrary, probabilistic symbols for its own material construction in the physical world. The inherent arbitrariness of symbols and measurements provides some spice, that is, some element of unpredictability, and is combined with the predictable menu of physical laws, resulting in life becoming both increasingly ordered and increasingly complex over time.

This distinction between subject and object is not just an interesting oddity. It begins at the level of physics in the distinction between the probability inherent in symbolic measurements and the certainty of material laws. The distinction is later exemplified in the difference between a genotype, the sequence of nucleotide symbols that make up an organism's DNA, and phenotype, its actual physical structure that those symbols prescribe. It travels with us up the evolutionary layers to the distinction between the mind and the brain.

For the past twenty-five hundred years, the discussions about thought and consciousness have focused on the human and, more recently, the fully evolved human brain. This has gotten us no further across the explanatory gap. It is time that we start exploring Howard Pattee's gap between living and non-living matter. If we determine how it was bridged, how life achieved semiotic closure, perhaps we can understand how to bridge the explanatory gap between the mind and brain. We even have support for this idea from William James! James, who went so far as to consider what he called the theory of *polyzoism*: "Every brain-cell has its own individual consciousness, which no other cell knows anything about, all individual consciousnesses being 'ejective' to each other."[29] The individual cell had some very rudimentary process that connected a subjective "self" with the objective mechanics. Semiotic closure, the link that spans the gap between living and non-living matter, is present within all cells. By realizing this and attempting to understand the processes involved there, we may begin to seek an understanding of consciousness from a different perspective and look for it in different places. I am not suggesting that single cells are conscious. I am suggesting that they may have some type of processing that is necessary or similar to the processing that results in conscious experience.

The explanatory gap has stumped us because the subjective experiences of the mind have resisted being reduced to neural firings of brain matter. They appear to be two irreducible complementary properties of a single system. We know that no matter how much is learned by objective external observers about the brain's structure, function, activities, and neural firings, the subject's experience of these firings is quite different from any observation of them. The details of the neurons' firing, or even that there are neurons firing, are not part of the subject's experience or intuitions. The objective workings of perception, thinking, and so forth are not available to the person doing the perceiving and thinking. As we discussed in the chapter on layering, those details are not necessary for the person and are hidden, abstracted from view. Furthermore, the function of the neurons can't be derived from their structure without any previous knowledge of their function, nor can their structure be derived from their function. Knowing all about one is not going to tell you anything about the other. They are two separate, irreducible layers with different protocols. Pattee believes that this is part and parcel of the complementarity principle, and that a single model cannot explain both objective structure and subjective function. The epistemic cut, the subject/object cut, is alive and well at the level of the human brain. Pattee states that "our models of living organisms will never eliminate the distinction between the self and the universe, because life began with this separation and evolution requires it."[30]

It should therefore not surprise us that two complementary modes of behavior, two levels of description, keep appearing in our thinking. The subject/object cut is present in all the great philosophical debates: random/predictable, experience/observation, individual/group, nurture/nature, and mind/brain. Pattee regards the two complementary modes as inescapable and necessary for any explanation that links the subjective and objective models of experience. The two models are inherent in life itself, were present from the beginning, and have been conserved by evolution. Pattee writes, "This is a universal and irreducible complementarity. Neither model can derive the other or be reduced to the other. By the same logic that a detailed objective model of a measuring device cannot produce a subject's measurement, so a detailed objective model of a material brain cannot produce a subject's thought."[31]

Ignoring one side of the gap will result in missing the link between the two sides. Linking the two requires acknowledging the dual and complementary nature of symbols. The link will consist of mechanisms that are describable by physics, yet the explanation may not prove warm and cuddly, not psychologically satisfying to anyone, neither to determinists nor to believers in spirits. It may be, just like quantum mechanics, something that nobody quite understands, way beyond our intuitions and imaginations. Feynman chided, "We are not to tell nature what she is going to be. That's what we found out. Every time we take a guess at how she's got to be and go and measure, she's clever. She always has a better imagination than we have and she finds a cleverer way to do it that we haven't thought of."[32]

9.

BUBBLING BROOKS AND
PERSONAL CONSCIOUSNESS

"It would be so nice if something
made sense for a change."

—Alice, in Lewis Carroll's *Alice's Adventures in Wonderland*

WE ALL SHARE this state we call consciousness, this awareness of our streaming thoughts, longings, emotions, and feelings about the world, others, and ourselves. It is not only omnipresent but personal, defining, and boundary setting. It defines the experience of living. The conscious self seems to ride above the physical brain and all its layers and modules. It seems that without it, we would be nothing but one of the automatons that Descartes observed in the gardens of Paris. A machine. So what would an explanation for consciousness possibly look like?

As you might guess, the elements of the puzzle that I think can lead us to some new thinking about the nature of conscious experience are the things we've covered in the preceding chapters: modules, layers, the principle of complementarity, and Howard Pattee's semiotic closure. These concepts will help us see how neural circuits are structures with a double life: they carry symbolic information, which is subject to arbitrary rules, yet they possess a material structure that is subject to the laws of physics. Combined, these perspectives tell the story of the brain. It is an organ, finely engineered by natural selection, organized into local modules whose functioning is accomplished in a layered architecture such that, for the most part, one set of modules doesn't know what all the others are doing. It is the story about how a bunch of hardworking small circuits exist in a coherent organization to produce a larger function, just as citizens, smart as they are, work in relative isolation to produce something like a society. The secret to understanding must include how the hardworking parts express themselves, moment to moment.

If gaps, modules, and layers are to help us understand how brains make minds, they must also account for some persistent facts about brains. Take a moment to fully consider this and its ramifications for your sense of self: a neurosurgeon can disconnect the two hemispheres of your brain and produce two minds in your single head—two minds with different contents at the same time, though with the same emotional drives and feelings. Next, remember: while brain damage can cause specific deficits, it is almost impossible to eliminate consciousness completely. And finally: while conscious experience seems unified and whole, it happens in concert, with multiple systems running parallel to one another, each separately spewing forth the results of its processing. Thus, while consciousness seems to be a cogently coherent, flawlessly edited film, it is instead a stream of single vignettes that surface like bubbles roiling up in a boiling pot of water, linked together by their occurrence through time. Consciousness is in constant change, a stream, and, as William James once said, "*No state once gone can recur and be identical with what it was before.*"[1] Let me set the stage for this idea.

Two Consciously Different Conscious Hemispheres!

I have to go back again to the very first scientific observation I ever made. It was on Case W.J., a man who had such severe epileptic seizures that he was only able to function normally about two days a week. A young neurosurgeon, Joseph Bogen, had done some extensive research and suggested that W.J. might benefit from a rarely performed surgery in which the large tract of nerves that connects the two cerebral hemispheres was cut. It had been done on a series of patients in Rochester, New York, twenty years earlier to control epilepsy. The surgery was successful in stopping or drastically reducing the seizures. Oddly, after having their brains cut in half, all those patients said they felt fine, and the only difference they noted was the loss of seizures.

W.J. was a World War II veteran and had fought many a battle. He weighed the odds of this one and agreed to the surgery. I was the young graduate student who had designed tests to run after he had undergone split-brain surgery to see what, if any, effects this surgery had wrought on his brain function. The expectation was that there would be none, since none had been found in the Rochester patients. W.J. was a warm, affable man, and the two hemispheres of his brain seemed to be working fine together, although they were no longer in direct communication. One of them talked, one of them didn't. Given the way the brain is wired, that meant the left, talking brain viewed the visual world to the right of a fixated point and the right, non-talking hemisphere viewed all the visual information to the left of the same point. Given this surgical state, I wondered: If I flashed a light over to the right, would W.J. say he saw it? Light should go to the left hemisphere, and the left hemisphere had speech; it should be easy. It was, and W.J. easily declared he saw it.

A bit later, I flashed the same light over to the left side of space and waited to see if he would say anything. He didn't. I pressed him and asked if he had seen anything, and he firmly said "No." Was he blind on that side? Or was that simple spot of light no longer communicated to the half brain that talked? Did the seemingly mute right brain know it had viewed a light? Was it conscious? What was going on?

It was later in that testing session that it became clear that the mute right hemisphere had indeed spotted the light, because the right hemisphere could easily and accurately point to it with W.J.'s left hand. It was the beginning observation that revealed that not only the brain but also the mind had been divided. It was the seed that led to sixty years of research on the nature of mind and its physical underpinnings. It was also the test that produced the most astounding observation of all. The left, talking brain didn't seem to miss the right brain, and vice versa. It didn't just not miss it—it didn't even remember it or the functions it had performed, as if the right hemisphere had never existed. For me, this phenomenon is the single most important fact students of mind/brain research must take into account.

How come the left brain doesn't complain about the fact that it is no longer conscious of things on the left side of space? Imagine having your brain disconnected. Imagine waking up in your hospital room the next morning and, as your surgeon walks in to see how you are faring, you only see the left half of her face. Don't you think you would notice that the right half is not there? The thing is, you would not. In fact, your left hemisphere wouldn't even be conscious of the fact that there is a left half of space. But this is the weird part: I spoke as if the new, split version of you were just your left hemisphere, and that is not true. You are also your right hemisphere. The new "yous" have two minds with two completely separate pools of perceptual and cognitive information. It is just that only one of the minds can readily speak. The other initially cannot. Perhaps, after many years, it will be able to produce a few words.

More crazy yet, in the early months after surgery, before the two hemispheres get used to sharing a single body, one can observe them in a tug-of-war. For example, there is a simple task in which one must arrange a small set of colored blocks to match a pattern shown on a card. The right hemisphere contains visuomotor specializations that make this task a walk in the park for the left hand. The left hemisphere, on the other hand, is incompetent for such a task. When a patient whose brain has recently been split attempts the task, the left hand immediately solves the puzzle; but when the right hand tries to attempt the task, the left hand starts to mess up the right hand's work, trying to horn in and complete the task. In one such test, we had to have the patient sit on his bossy left hand to allow the right to attempt the task, which it never could accomplish! It was beyond the capabilities of the left hemisphere.

When communication between the hemispheres is lost, each is unaware of the other's knowledge and each functions independently based on the information it receives. Both sides of the brain try to complete the task independently, resulting in the tug-of-war. By this simple task, the illusion of a unified consciousness is exposed. Clearly, if consciousness arose from a single location, then a split-brain patient would be unable to have two simultaneous experiences!

It gets better. We have all seen the film of the simple illusion that occurs when two balls seemingly hit each other, and, after the faux collision, the supposedly impacted ball takes off. In psychological parlance this is called the *Michotte launching effect*, after the Belgian psychologist Albert Michotte, who first invented the illusion to investigate how we perceive and infer causality. Since the first ball stopped next to the second ball and didn't actually hit it, there is in fact no physical event that happened to cause the second ball to take off. But that is not how we all view the experience. Ball A hit ball B and it took off, period. There was causality!

So how do split-brain patients view this simple task? Does the left brain, with the flavor of consciousness it possesses, see it the same way the right hemisphere might see it? An experiment to test this was first initiated by Matthew Roser, originally a student from New Zealand, who was then in my laboratory at Dartmouth and is now residing at Plymouth University in England.[2] An exceptionally talented scientist, Matt, along with other colleagues, examined how each half brain, working alone and disconnected from the other, viewed the seemingly colliding balls. The results were astonishing. The right hemisphere instantly grasped the illusion, while the left brain did not. This was established by a second experiment, in which the distance between the balls at the point of the faux collision was slightly increased, or the amount of time it took for the second ball to move was increased. In these circumstances, the illusion disappears for the right hemisphere and never occurs. The left brain, the one that does the heavy lifting on all matters cognitive, simply didn't ever see through the illusion under any conditions. Interestingly, the left brain does see relationships that the right seems unable to grasp. In the context of these tests, it was the left brain that picked up on another problem, requiring logical inference, which the right brain was unable to do. In short, the direct *perception* of causality was something the right hemisphere could do, but the ability to *infer* causality was only in the left hemisphere's bag of tricks.

When we consider how the normally connected brain perceives these two kinds of tasks, it suddenly becomes clear that when seeing the illusion, it is the right hemisphere that has the neural equipment to apprehend it. And, when tackling a logical inference task, it is the left hemisphere that is processing the information. So, in a connected brain, at time A, when the right hemisphere is seeing a launching ball test it is saying "Hey, ball A just hit ball B," but at time B, when it is looking at a logical inference kind of test it is the left hemisphere, not the right, that is able to apprehend it. It is kind of like the arcade game Whac-a-Mole. We are aware, which is to say conscious, of the processed information as it pops up after being processed in a particular hemisphere. But is that because each neural process activates a "make-it-conscious network" (which would have to be present in each hemisphere), or does each process in and of itself possess the neural capacity to appear conscious?

Tiny Bubbles

I am in the latter camp. Thinking about this question led me to use the metaphor of bubbling water as a way to conceptualize how our consciousness unfolds. Consciousness is not the product of a special network that enables all of our mental events to be conscious. Instead, each mental event is managed by brain modules that possess the capacity to make us conscious of the results of their processing. The results bubble up from various modules like bubbles in a boiling pot of water. Bubble after bubble, each the end result of a module's or a group of modules' processing, pops up and bursts forth for a moment, only to be replaced by others in a constant dynamic motion. Those single bursts of processing parade one after another, seamlessly linked by time. (This metaphor is limited to bubbles roiling up at a rate of twelve frames a second or faster; or consider a cartoon flip book, where the faster we snap the pages, the more continuous the movements of the characters appear.)

Sir Charles Sherrington, the doyen of modern neuroscience, had a related notion when he observed:

> How far is the mind a collection of quasi-independent perceptual
> minds integrated psychically in large measure by temporal concurrence
> of experience? Its separate reserves of sub-perceptual and perceptual
> brain, if we may so speak, could account for the slightness of the
> mental impairment following on some brain injuries. . . . Simple
> contemporaneity can conjoin much.[3]

It's difficult to get our heads around the idea that each bubble has its own capacity to evoke that feeling of being conscious; it rubs up against our own intuitions about the holistic nature of our personal consciousness. What are we and our intuitions missing? We are missing the illusion part, the part we humans (with our powerful left hemisphere inference mechanism) are so good at missing. We aren't actually missing the illusion; rather, we are missing the fact that our smoothly flowing consciousness is itself an illusion. In reality it is made up of cognitive bubbles linked with subcortical "feeling" bubbles, stitched together by our brain in time.

Background for Bubbles

There is a classic observation that rings true across all of biology. The observation concerns whether organisms learn and take instruction from the environment or whether the reactions that organisms have to environmental stimuli are managed by systems already built into the organism. The "selection versus instruction" debate has raged for years and has especially caught the limelight in the field of immunology. Put simply, when something foreign enters the body and there is an immune response to it, are the antibodies formed then and there around the foreign body, and do they then multiply (instruction)? Or does the antibody already exist, and is the immune response time the time it takes to find the preexisting antibody (selection) and jerk it into action? In the previous century, biology learned it is the latter situation, a finding that illustrates that a whole lot of stuff comes with the package—standard equipment for our bodies and brains.

Niels Jerne, the Danish immunologist, proposed in 1967 something rather startling at the time: what is true for the immune system is probably also true for the brain. He suggested that preexisting circuits in the brain are selected by the environment and are applied during what we might think is a "learning" situation.[4] With this strong naturalistic view, learning is simply the time it takes for the brain to sort through its gazillion circuits to find the one more appropriate for the challenge at hand.

While this crucial debate rages on, no one would doubt that there are neural circuits for specific functions that come as "standard equipment" with our brains. For example, babies as young as six months have demonstrated that they already possess the ability to make causal inferences.[5] And subcortical circuits make their processing apparent as soon as that newborn cries for its first meal. In the bubble metaphor, bubbles are the end result of the processing of those circuits that are constantly in play to cope with and manage the endless challenges of the environment. Those processes are both cortical and subcortical. Let's look at another telling experiment before we get to the subcortical bubbles.

Bats in the Belfry

For better or for worse, Thomas Nagel's infamous question "What is it like to be a bat?"[6] has stirred the philosophically minded for forty years. Actually, the question should be "What kind of bubbles does a bat have?" That is, what are the contents of a bat's consciousness? We will probably never fully experience bat-brand consciousness, but we can observe the contents of a lonely single hemisphere in a human brain. The brain is full of bubbles, and when a brain is split, each half has its own set of bubbles that boil up. Since we now know that each half brain has some unique bubbles, might it be that each hemisphere harbors a different kind of overall conscious experience? To get a feel for this, consider what bubbles you don't have. For instance, I can tell you I don't have the abstract math bubble; therefore, I can tell you I feel frustrated when equations start popping up in lectures. Though I wish I could, I cannot tell you what it is like to grasp highly abstract math, but I bet it would be cool!

Rebecca Saxe, at MIT, has discovered in some intriguing studies that there is special human brain hardware in the right half brain that appears specialized for determining what the intentions of another person might be.[7] When we interact with others, we are constantly and reflexively making assessments of their mental state and their intent in all of their actions. It is virtually automatic. It appears that children with autism lack this capacity to a large extent and, as a result, find social interactions difficult. As I discussed earlier, in formal psychological parlance it is called having a theory of mind. Saxe, using modern brain imaging techniques, discovered a brain area in the right hemisphere responsible for this capacity. As you might guess, this observation raises a new question. The Saxe finding would suggest that perhaps the left hemispheres of split-brain patients may not have access to the module that adds theory of mind to our cognition. What would a left hemisphere be like that didn't have access to that capacity?

Michael Miller, my former student and now colleague, and Walter Sinnott-Armstrong, the distinguished philosopher, teamed up to examine the implications of the Saxe finding for split-brain patients.[8] They wanted to determine if one separated hemisphere might evaluate moral issues differently than the other. Again, in a split-brain human, Saxe's work suggests one hemisphere (the right) would have the module that considers the mind and intentions of others, while the other hemisphere (the left) would not. Upon separation, would the left hemisphere act differently, since it no longer possessed a module that evaluated the mental states and intentions of others?

Moral philosophers like to approach moral dilemmas as having either a deontological or a utilitarian nature. In plain English that means, "Do we solve the dilemma by considering what is inherently right, what our moral duty is, or does the solution reside in maximizing collective good?" There are many ways of phrasing this dichotomy and many ways to reveal whether someone is more deontological in their thinking or more utilitarian. In a series of cleverly devised tests, patients were told stories in which the main person did something evil but the outcome was, nonetheless, un-harmful: If a secretary wants to bump off her boss and intends to add poison to his coffee, but unknown to her, it actually is sugar, he drinks it, and he is fine, was that permissible? Or the story was about a person doing something that seemed innocent to them but proved to be fatal to someone else: If a secretary believes that she is adding sugar to her boss's coffee, but it actually is poison accidentally left there by a chemist, and her boss drinks it and dies, was that a permissible action? The patient, upon hearing the full stories, had to simply judge whether the act the person did was "permissible" or "forbidden."

Needless to say, most people judge an example with mal-intent as forbidden no matter what the outcome. Most people in this sense are deontological. Most people would judge an action by someone who had no malicious intentions to be permissible (although not always), even though it sometimes can end in tragedy. Split-brain patients act in a unique way. It appeared the left, speaking hemisphere initially offered a utilitarian response to all scenarios. Thus, if an act had mal-intent but no harm came from it, it was judged as "permissible." And if an act did not have mal-intent, but resulted in harm, it was judged "forbidden." Given the clarity of the stories used, this was a jarring result. What is going on? The disconnected left hemisphere is unable to take into account the intent of the person in the stories, acting as if it didn't have a theory of mind.

Second, the patients would then frequently give spontaneous explanations as to why they had chosen the utilitarian result over the clearly deontological choice. It seemed they "felt" their judgments were not exactly copacetic, and they often rationalized their judgment without any prompting. Remember, the left hemisphere does have its interpreter, the module that tries to explain both the behavior it observes pouring out of the body and the emotions it feels. Keep in mind that an emotional reaction to something experienced by one side of the brain is felt by both. If the emotion was a result of the right brain's experience, the left brain has no information about why it is feeling the emotion it is, but explains it anyway. So, when the right brain heard the left brain's answer (even with limited language abilities there is still some comprehension in the right hemisphere), it was just as shocked as we were, resulting in an emotional reaction that didn't match what the left hemisphere considered to be a reasonable answer. With the stage set for a major conflict like this, it was not surprising that the special module in the left hemisphere (the "interpreter" module—the one that is ever ready to explain away behaviors produced from the silent disconnected right hemisphere) jumped in and tried to explain what was going on. For example, in one scenario a waitress served sesame seeds to a customer while falsely believing that the seeds would cause a harmful allergic reaction. Patient J.W. judged the waitress's action "permissible." After a few moments, he spontaneously offered, "Sesame seeds are tiny little things. They don't hurt nobody."

In my metaphor, a bubble is the end result of the processing of a module or group of modules in a layered architecture. The special module that evaluates the intent of others in split-brain patients is disconnected and isolated from the speaking left hemisphere. As a consequence, the result of its processing doesn't bubble up to contribute to or battle for dominance in the left hemisphere's decision-making process. It can't be part of that bubbling process if it is not physically located in the left hemisphere among the bubbles having access to language and speech. So, eerily, the knowledge of the intent of the other is absent. Yet bubbles from midbrain emotional processing do make it to both hemispheres. It is only when the right hemisphere hears the left hemisphere's response, resulting in an emotional feeling felt by both hemispheres, that a mismatch is identified by the left hemisphere. That sets the process of justification in motion. The left hemisphere also has a lifetime of memories stored about the moral norms of the culture it has grown up in and can use these for justifications.

We are discovering that the subtleties of our psychological lives are being managed by specific modules in our brains. Again, the left brain, which benefits from modules that enable abstract thinking, verbal coding, and much more, does not have the module to take into account the intentions of others. Yet it has a powerful inference ability. If the result is good, it infers the means were okay. Thus, if the result is fine, the act is permissible. If the end is bad, the act is not permissible. What is the best for the most is okay. The uncanny and almost surreal aspect of these findings is the possibility that if the proper module that enables one to think about others is missing, one can't seem to learn it.

What Is It Like to Be a Right Hemisphere?

What if you suddenly lost your left hemisphere? What would your conscious experience be? Remember that you will not actually notice that anything is different, because you will not miss the left hemisphere's modules. Really. Having lost the speech center, communication and comprehension abilities would take a nosedive. The right hemisphere would only be able to manage limited comprehension and vocabulary. But one of the biggest contributors to a changed experience would be the loss of your ability to make inferences. You constantly use this left-hemisphere ability, and lacking it would plop you into a very different experience of the world. Though you would know that others had intentions, beliefs, and desires, and you could attempt to guess what they might be, you would not be able to infer cause and effect. You would not be able to infer *why* someone is angry or believes as they do. Many of your social encounters will probably result in misunderstandings and frustration on both sides. But the loss of your inference ability is not limited to the social world. You would not infer causality at all. Not only do you not infer that your neighbor is angry because you left the gate open and her dog got out, you don't infer that the dog got out because you left the gate open. You don't infer that the car won't start because you left the radio on.

While you would be good at spatial relations, you would not grasp the causes and effects described by physics. You will not infer any unobserved causal forces, whether they be gravitational or spiritual. For example, you would not infer that a ball moved because a force was transferred to it when it was hit by another, yet because of your inability to draw inferences, you would do better in Vegas at the gaming tables. You would bet with the house and not try to infer any causal relationship between winning and losing other than chance. No lucky tie or socks or tilt of the head. You would not string out some cockamamy story about why you did something or felt some way, not because you aren't capable of language, but again because you don't infer cause and effect. You won't be a hypocrite and rationalize your actions. You would also not infer the gist of anything, but would take everything literally. You would have no understanding of metaphors or abstract ideas. Without inference you would be free of prejudice, yet not inferring cause and effect would make learning more difficult. What processing comes bubbling up in your separate hemispheres determines what the contents of that hemisphere's conscious experience will be.

Feelings

We have all experienced the longing to return to a favorite spot in the world, a place we once relished and remember so fondly that we had to go back and recapture the moment. And yet when we go back to visit the second time, it is not quite the same. The feelings are different. It is not that they are bad or good. It is that they are different.

My wife and I recently returned to Ravello, a place we had found magical in the past. It still had its natural beauty. It still had its history, culture, and, best of all, its people. Yet it felt different to us. It didn't seem to match our feelings about the place, rooted in our previous experience. Were we misremembering something or were we "feeling" different about life in general, and were those different feelings coloring our current experience?

Our commonsense idea about past experiences is that each has a feeling to it that is specified for the particular event in question. The actual feeling itself is tied to the actual experience, and we expect we can recapture the feeling by replicating the experience. You felt happy and excited when you vacationed in place x; if you return, then you will again feel happy and excited. This may be why people buy time-share condos. Their first visit is great, but the next . . . and the next? Repeat the event and we expect the original feeling will come along automatically. I am suggesting that it doesn't. This is not how to think about it.

That magical moment that we relished in the past is now stored as information about that past event. It is placed in memory in ways we still don't understand, but it is symbolic information, cold and formal, just as DNA is symbolic information—and, just like DNA, it has a physical structure. Contained within that informational structure may be the fact that the event was associated with positive feelings, but the feelings themselves are not stored, just information about them. As a memory bubble pops up, so do bubbles spewing forth our *current* feeling about it. Feelings are from another system with its own processing bubbles, separate from the memory system. In short, I am suggesting that uncoupling the emotional dimension reveals that different systems are converging in time to produce a feeling about a memory. For an analogy, it's like the musical soundtrack of a dramatic movie scene. They are separate, but when they are put together, the soundtrack adds emotion. As one bubble quickly passes to the next, we have the illusion of feeling about the remembered event.

So we have memory bubbles and feeling bubbles. When we think about that past moment, our feelings about it are not actually the feelings of the past moment; rather, they are our current feelings, which we map onto the past moment. Usually they, too, are positive, but the feelings themselves are not bolted to the actual past memories. This is more obvious when today's feeling is the opposite of the original feeling. For example, say you are recalling the moment when you received the results of an exam you failed in college. Your memory tells you that you felt bad at the time, right? But say that, because of this failure, you marched yourself into the professor's office and asked for help, which led to you actually becoming engrossed in the subject, which led to a career in the field. When you look back at the event of the bad grade, the current you knows where it led and now can think of it in only positive terms. Intellectually you remember that you felt bad, but you just can't reincarnate those feelings. Perhaps more resistant to changes in perspective are feelings of embarrassment. You may still blush thinking of some past social transgression. Yet sometimes you can only laugh and shake your head at what embarrassed the younger (teenage) you: hiding on the floor of the backseat at the drive-in so no one would see you with your parents. Did I really do that? No blushing involved now, or maybe blushing for an entirely different reason.

The modules that produce *raw* feelings are different from the ones that produce thoughts, memory, decisions, and so forth. What we are feeling at a moment in time during an event becomes an aspect of the memory of the event, a dimension, a piece of information we can label and neurally code and include in our memory.

The actual feeling, however, comes from another, totally independent contractor in the brain. On a revisit, it can be on different neurological settings from the previous visit. At both times, it was those neurological settings that were driving your feeling about the visit. The first visit involves the challenge of the unknown, curiosity, adventure, often youth, and sometimes adrenaline. On a revisit, you are returning older, with more life experience under your belt, to familiar territory; it is not so challenging, your curiosity is not so piqued, you know more or less what to expect, you have less adrenaline pumping. You feel different, not necessarily worse or better, just different. This is true not just for returning to locales but also for trying to recapture any past experience. Neil Young describes this well: "I still try to be that way, but, you know, I am not twenty-one or twenty-two. . . . I am not sure that I could re-create that feeling, it has to do with how old I was, what was happening in the world, what I had just done, what I wanted to do next, who I was living with, who my friends were, what the weather was."[9]

Sentio Ergo Sum

A few years back, Steven Pinker observed: "Something about the topic of consciousness makes people, like the White Queen in *Through the Looking Glass*, believe six impossible things before breakfast. Could most animals really be *unconscious*—sleepwalkers, zombies, automata, out cold? Hath not a dog senses, affections, passions? If you prick them, do they not feel pain?"[10] Jaak Panksepp agreed with Pinker that to deny that is to believe the impossible, and blamed Descartes, thinking he was way out of line in denying animals consciousness. Not only that, he thought Descartes would have saved us all a lot of trouble if, when he asked "What is this 'I' that 'I' know?" he had just replied, "I feel, therefore I am," and left cognition out of the subjective-experience equation. Panksepp would have agreed with Pattee that most everyone since Descartes has been looking way too high in the evolutionary tree for how neural systems manage to produce subjective affective experience.[11]

Panksepp proposed that subjective affective experience arose when the evolutionarily old system of emotions linked up with a primitive type of neural "body map" that delimits the "self" from the external world.[12] To form the body map would only require sensations from inside and outside the body to be tacked onto sheets of related neurons in the brain. He argued then that the two things necessary for subjective experience are information about the internal and external states of the agent (recorded symbolically), and the construction of an integrated neural simulation of the agent in space: a quick and dirty model, built from the firings of neurons. Information and construction, the same complementarity we saw in DNA. Higher cognition or knowing that you have a "self" (a.k.a. "self-awareness") is not part of the original recipe. To move through the environment safely and efficiently, to eat when you are hungry, and so forth, you do not need to know that you are self-aware, but you do need to know the boundaries of your body in relation to the space that surrounds it. If you didn't know that, you would be forever bumping into things and misjudging everything from fitting into a cozy den to jumping from one rock to another. And you also need to have the motivation to act in ways that promote survival and reproduction. That is, you don't need bubbles roiling up from a highly evolved cortex to be aware of a subjective experience. Descartes only needed signals from the subcortex to feel he was an "I"; he didn't need to think about it.

In fact, even an insect needs to have information about its body in space in order to move safely and effectively. I have been outmaneuvered by many a fly, wielding my swatter to no effect. The biologist Andrew Barron and philosopher of neuroscience Colin Klein, both from Australia's Macquarie University, began prowling around the world of insect brains and found that structures known as the *central complex* performed functions analogous to those of parts of the vertebrate midbrain, generating "a unified spatial model of the state and position of the insect in the environment."[13] That is, functions that locate them in space are present in the brains of cockroaches and crickets, locusts and butterflies, fruit flies and honeybees, just as they are in the midbrain of vertebrates. In short, the task of ordering the plethora of systems needed to perform an action is an extensive biological feature of complex organisms. Bugs have it, we have it. Barron and Klein, agreeing with Panksepp and Merker, conclude, "This integrated and egocentric representation of the world from the animal's perspective is sufficient for subjective experience." They also think this awareness of the body in space is sufficient for the insects they have studied, suggesting that they, too, have a form of subjective experience, which was present in a common ancestor of both vertebrates and invertebrates back in the Cambrian explosion, about 550 million years ago (MYA).

They are not alone. The neurobiologists Nicholas Strausfeld, from the University of Arizona, and Frank Hirth, from King's College London, are also venturing out on their own limb of evolution's tree. They did a massive review of the anatomical, developmental, behavioral, and genetic characteristics of vertebrate basal ganglia and compared them with those of the central complex of arthropods (of which insects are but a branch). They found so many similarities that they were prompted to conclude that the arthropod central complex and vertebrate basal ganglia circuitries shared a common ancestry.[14] In fact, the two are derived from an evolutionarily conserved genetic program. That is, the circuits essential to behavioral choice originated very early in evolution. The ancestor that we vertebrates share with the arthropods was already scampering around with such a circuit, with bubbles of processed information about its location and sensations popping up to guide its actions. Strausfeld and Hirth even suggest that common ancestor's brain gear was sufficient to support phenomenal experience. While their view may be overdrawn, their work does alert us to how deeply conserved the basic mechanisms we unearth in humans may be. This is the beauty of evolutionary and comparative research. Aspects of our own mental life that we assume to be unique constructions of the human brain have, in fact, been around for a very long time and are mainly elaborated upon by human brains.

Methodic or Chaotic Bubbles?

Our conscious experience is a continuous, smoothly running flow of thoughts and sensations. How can this come about with all these bubbles battling it out for precedence? Do the bubbles burst willy-nilly, or are they the product of a dynamic control system? Is there a control layer giving some bubbles the nod and quashing others?

One way a module's processing is controlled is by the input it receives.

Let's say you bite into a chocolate truffle made with unsweetened chocolate. No afferent nerve, that is, one that comes from the periphery to the brain, is activated by taste cells that detect sweetness, no module that processes such sensations is activated, so you do not taste anything sweet, and no sweet information is getting processed. Instead, the taste cells that detect bitterness are activated, and your mouth is flooded with a bitter taste. Bitter information is getting processed. Swap out that unsweetened chocolate for an identical-looking milk chocolate one, and that sweet module is up and running, too, bubbling sweetness into your awareness like a house on fire, outcompeting bitterness by a landslide. It is as if bitterness is now a distant memory and the moment is owned by sweetness, until the next bubble arrives. No cognition is required here. Somehow, that sweet signal won the bubblefest. Did it get some additional help?

Some form of selective signal enhancement has been found in animals ranging from crabs[15] to birds[16] to primates,[17] suggesting it is an ability shared with our last common ancestor, which was roaming around about 550 MYA. The earliest manifestation of this "data control" ability was a rudimentary form of attention, a process that helped manage the onslaught of sensory information besieging an exploring hunk of cells. A signal-enhancing process appeared early in evolution to help organisms sort out which of all the stimuli bombarding them might be more relevant for survival (better to give priority to information about an approaching threat, meal, or mate than to much else) and has been conserved across all life forms, stemming from this early organism.

Steven Wiederman and David O'Carroll at the University of Adelaide in Australia discovered that in the modern-day dragonfly brain there is a single visual neuron that selects one particular prey-like visual target and follows it, completely ignoring another.[18] This is interesting not only because it demonstrates that dragonflies have a form of competitive selection, which is required for visual attention, but also because a selective attention process is accomplished by a single cell. In vertebrates, selective signal enhancement evolved into what we now call attention, a sophisticated mechanism that controls incoming data and, through that, our minds. Thus, we may be engrossed watching an exciting movie, but one squeal of a fire alarm and our "data control" system immediately enhances that screeching input, jerking our minds away from the movie and into action.

Yet, although selective signal enhancement has some control over our minds, our minds also control our attention to some extent. What I mean is that attention has two components, bottom-up and top-down. If you are meeting a blind date with a red flower in her hair, your eyes will be swiveling around looking at hairdos and not much else. Your attention is top-down, being guided by your plan to find the unknown date. Bottom-up attention will only get you so far in the survival game. Those organisms that developed top-down attention had an edge, and it became de rigueur. This top-down attention ability is highly developed in both birds and mammals, whose common ancestor was cruising around 350 million years ago, so this ability is at least that old, but evolutionarily newer than bottom-up attention. An add-on app: a new layer.

Once again Pattee makes a useful suggestion about how such a layer could evolve. The failure of one layer is the basic force or condition for a new one to arise:

> When a system fails to have a representation or a description to handle a particular situation, it leaves a power vacuum so to speak, or a decision vacuum. I would call it a kind of instability, when a decision needs to be made and there is no decision procedure. One then has ambiguity, and small causes can have large effects. This is, in effect, a crisis in the system, and there can arise then a new type of behavior.[19]

So as systems became more complex, some sort of control layer was needed to manage the plethora of independent stimuli and resultant behavior. While stimulus enhancement is great, a control layer was needed to somehow orchestrate the squawking of all the modules.

A Famous Control Layer Malfunction

As I discussed earlier, people with right parietal lobe lesions can experience neglect of the left side of the visual field even in their imaginations and memories. This famous observation was originally made by the brilliant Italian neuropsychologist Edoardo Bisiach.[20] In a nutshell, he asked patients to describe from memory the Piazza del Duomo in Milan from two different perspectives. It is a beautiful site, with buildings of a particular kind lining each side of the piazza and the grand cathedral at the end. Everybody in Milan can imagine it with ease.

When asked to describe the Duomo as if they were standing in front of it and looking straight at it, the patients easily did so, but they only described the buildings lining the right (north) side of the square. They left out any description of the buildings on the left (south) side. Then Bisiach followed up by asking them to imagine themselves turned around 180 degrees and to describe the square from the perspective of standing on the front steps of the Duomo. Now the very same patients easily described the buildings on the right (south) side looking out from the Duomo and never mentioned the very buildings on the left (north) side that they had just described when imagining facing in the opposite direction!

This dramatic clinical example reveals the existence of two completely different sets of modules. Clearly, the modules generating a mental image remain intact, and all the information is present, but those modules are controlled by an additional module that evaluates from which side of space the image will be reported, and this module is malfunctioning. Years later, the neurologist Denise Barbut and I were able to examine a case at New York Hospital with similar damage to the right parietal lobe and replicate Bisiach's findings.[21]

Consciousness Enriched by Evolution

When it comes to consciousness, we seem to have forgotten the fact that our brains evolved by adding complexity. Over the course of evolution, modules and layers have been annexed over time to solve one perturbation after another, changing and increasing the content of our conscious experiences along the way. Each layer has its own independent rules for processing and passes its handiwork, its processing bubble, on to the next. While the processing within a module, going from layer to layer, may be serial, multiple modules are running in parallel, bubbles roiling up from each, coming to a final realization.

In the bubble analogy, the results of processing from various modules burst into our conscious awareness from one snippet of time to the next. Most likely, one bubble is boosted into the limelight by a control layer with a protocol made up of arbitrary rules, rules that have been selected because they have provided the content of consciousness with the most reliable, apt information for the situation being faced. A better and more reliable rule comes along, and the protocol can be changed. For example, consider a belief bubble that comes popping up to consciousness. Say you believe that saturated fat is unhealthy and eating it is making you gain weight. You believe that you should replace those calories with carbohydrates. You believe this because it is what you have been told by public health officials, nutritionists, and your doctor. When you are grocery shopping, bubbles are popping up, guiding you to stay clear of saturated fat. But then you make an observation. Your experience does not support the official's claims. The more fat you cut out of your diet and replace with carbohydrates, the more weight you gain. Then you read a tome in which the author reviews and evaluates all the research behind this claim, and finds that not only does most of the research not meet the standards of good science, but the small fraction that does fails to support the claim. In fact, it suggests the opposite is true.[22] You end up convinced. You buy butter and cream and eat it. You lose weight. At the grocery store the cream bubbles take over. A better and more reliable rule has come along and the protocol has changed. It makes for a better cup of coffee, too.

Importantly, gaps, modules, and layers can help us understand the behavior we observe in individuals who suffer brain lesions. If we lose some neural tissue that processes particular aspects of information, then that information is no longer part of our repertoire of bubbles and no longer provides content to our conscious experience. The same is true when the right hemisphere is separated from the left: neither hemisphere has bubbles from its opposite member popping up to enrich its conscious experience, leaving the conscious experience of each hemisphere impoverished.

10.

CONSCIOUSNESS IS
AN INSTINCT

One man's "magic" is another
man's engineering.

—Robert A. Heinlein

ACK IN MY YOUTH, right when I was beginning my graduate studies at Caltech, I became friends with the political philosopher Willmoore Kendall. He was the original disruptive personality, so exasperating that the Yale administration had paid him a large sum of money to resign. Kendall rocked all of their assumptions on just about everything and then headed out west. The West was not unfamiliar to him, having been born in Konawa, Oklahoma, to a blind minister. Ultimately, after Yale, he settled down in a small Jesuit school in Dallas. His appetite for life had no bounds, and it came from a man with a sure-footed ego. On the day JFK was shot in Dallas, I got a call from him. Kendall declared, "I never before have been at a lunch where the president of the United States spoke. I should have known something would happen."

Kendall kept nudging my thinking, as he saw me as a victim of modern reductionist thought gone wild. I was, and to this day remain, committed to the idea that physical mechanism can and will explain almost everything. By and large, when philosophers start parsing foundational thinking, laboratory scientists' eyes glaze over. Kendall was battling someone who wasn't up to the deep fight and indeed was barely aware that there were issues. Since I kept lending him my apartment during his day visits to Pasadena, he decided he should reciprocate and provide me with a larger education. He directed me to read Michael Polanyi's classic book *Personal Knowledge*, which grew out of Polanyi's 1951 Gifford Lectures. I did. Ever since, it has hovered in the recesses of my mind and in the recesses of my bookshelf, dragged back and forth across the country umpteen times.

Polanyi, Kendall used to tell me, was a true polymath. During sick leave while serving as a physician on the Serbian front in 1916, he wrote his chemistry PhD thesis. Although he held a chair in physical chemistry at the University of Manchester, his wide-ranging interests in economics, politics, and philosophy resulted in the university creating a chair for him in social science. "You know," Kendall commented, "he answers his correspondence each day in twelve different languages." A giant in his time, Polanyi, though based in England, was a regular visiting lecturer to the University of Chicago. It was reading his book that, for me, raised the thorny problem that knowing the parts of something doesn't always tell you what the whole might be. There is something else going on, something missing, and I think this framing gave rise to what I will call the Chicago school of thought, that is, a particular perspective on brain processes.

That something was missing was the result of the machine metaphor, which had been around since Descartes and had been swallowed hook, line, and sinker by biologists. The Chicago scientists had realized that the traditional deterministic classical machine analogy for life is exactly backward. Brains aren't like machines; machines are like brains with something missing. Polanyi pointed out that humans evolved through natural selection, whereas machines are made by humans. They exist only as the product of highly evolved living matter, and are the end product of evolution, not the beginning.

The Chicago perspective also launched the idea that the origin of life depended on two complementary modes of description, not just the description provided by classical physics that works so well for machines. Pattee summed it up thus: "Life itself could not exist if it depended on such classical descriptions or on performing its own internal recording processes in this classical way."[1] This followed up on Rosen, who had really shaken things up earlier by asking, "Why could it not be that the 'universals' of physics are only so on a small and special (if inordinately prominent) class of material systems, a class to which organisms are too *general* to belong? What if physics is the particular, and biology the general, instead of the other way around?"[2]

The battle against pure reductionism started with Polanyi and the University of Chicago professor Nicolas Rashevsky, the father of mathematical biophysics and theoretical biology, who was an unlikely soldier to battle against reductionism. He had initially taken up the problem of establishing the material basis of basic biological phenomena in general, after being outraged at a party when a biologist told him that nobody knew how cells divided, and it was something no one could know because it was biology, outside the pale of physics. After a stunning amount of work on the problem in the 1930s and '40s, he was growing uneasy. As Rosen, his student, describes, "He had asked himself the basic question 'What is life?' and approached it from a viewpoint tacitly as reductionistic as any of today's molecular biologists. The trouble was that, by dealing with individual functions of organisms, and capturing these aspects in separate models and formalisms, he had somehow lost the organisms themselves and could not get them back."[3] He came to the realization that "no collection of separate *descriptions* (i.e., *models*) of organisms, however comprehensive, could be pasted together to capture the organism itself. . . . Some new *principle* was needed if this purpose was to be accomplished."[4] Rashevsky dubbed that pursuit of the new principle *relational biology.* In many ways my mentor Roger Sperry, who was trained at the University of Chicago, took up the search as well. These ideas, as we have seen, also deeply influenced Howard Pattee, who receives the credit for bringing this line of reasoning into the present.

The gnawing message from the early Chicago story was: there is something else that needs to be accounted for when considering an organism. Mechanistic thinking is fine and teaches us all about the parts, the layers of automatic processes that are tirelessly at work in any organism in order to allow it to exist. But there is something else, another factor that needs to be understood, and that something is not a spook in the system. It is the system, the organism itself, that can modulate the lower layers that produce it. It is what answers the question "What is life?"

As Rosen argues, science always inserts a surrogate (a model) for the actual thing it is trying to study. With the surrogate, scientists can use all the methods of reductionist science and figure out how the parts work. The assumption is that the surrogate can substitute for the real thing. But when they go back to the real thing after working on the surrogate and try to plug in their findings, they usually fall short. For example, studying the pancreas alone in a dish underneath a microscope or in a test tube is one thing. It can teach us how it functions locally. But unless you study it all connected up with the body, you are not going to understand its real function, or how it works in concert with and is modulated by a distant system—in this case, a piece of the intestine. The fact that the functions of the pancreas and the intestine are entwined was not stumbled on until surgeons started doing lap-band surgeries for obesity and observed diabetes disappearing overnight. For neuroscience, the surrogate that had been substituted for the brain was "a machine." And by thinking of it as a machine, they were going to miss the whole idea of complementarity and what that buys us for understanding how the brain does its tricks.

Sperry put it differently. When he suggested our mental capacities were real entities and part of the causal chain of events that lead to behavior, the reductionists went crazy, as we learned in chapter 3. Yet he didn't see mental events, such as thoughts, as nonphysical events or spooks in the system, either. He saw mental events as the product of the configurational properties of the underlying neural circuitry. That underlying circuitry has both a physical and a symbolic structure. It controls what it is constructing, a mental event—Pattee's physical symbols controlling construction. In short, he had the organism itself playing a role in its own destiny. From this perspective, even knowing every possible thing about the current state of your brain—its initial conditions—would not allow you to predict how future mental states may have a top-down effect on your bottom-up processing. Those initial conditions are not going to tell us what, where, and with whom you are going to eat dinner a year from Thursday. Knowing everything about the state of a newborn's brain is not going to allow you to know what that child will be doing on a Tuesday afternoon forty-five years later, as the most determined of the determinists believe. In fact, that extreme determinism is almost as silly as the belief revealed by the Schrödinger cat problem.

Looking Forward

In our sketch of the history of human thinking and research on the problem of consciousness, we have seen a lot of equivocations. It was only after Descartes and the birth of the idea that "the brain is a machine that can be understood by taking it apart" (the sine qua non of the scientific approach to anything) that the ironclad devotion to reductionism firmly took hold, and it remains the dominant idea in neuroscience today. Again, the Chicago school, as I have come to call it, puts the brakes on that and has pointed the way to another formulation, which takes into account the evolutionary nature of the organism and the fact that machines are by-products of human brains— brains are not the by-products of machines. There is something different about living matter. Put bluntly, it is the fact that it is not solely at the beck and call of classical physical interactions, but has an innate arbitrariness conferred on it by physical, yet arbitrary, symbolic information residing on the sunny side of the *Schnitt*.

When the early results of split-brain research became known and established, the persistent question became: So what does it teach us about consciousness? As the famous experimental psychologist William Estes quipped to me after I was introduced to him as the man who discovered the split-brain phenomenon, "Great, now we have two things we don't understand." Yet it was that very puzzle that has stayed with me, just like the Polanyi articulation that the parts list doesn't tell you how something works. Both realities, the parts list and how those parts work together to produce its function, demand a more complex explanation about how these facts illuminate the problem of consciousness.

Over the past thirty years, billions of dollars have been invested in the study of the role of various brain regions and how they are connected. Yet localization will not yield a comprehensive explanation of consciousness, even though modern brain studies tell us that specific anatomical areas are related to various mental capacities. Although these studies add to the plethora of facts known about the brain, they do not and will not provide explanations of the processes the brain performs, which result in, among other things, consciousness. While the structure-function approach provides insightful knowledge about how the brain compartmentalizes its many specializations, it fails to adequately explain how electrochemical reactions are transformed into life experiences. We have seen that structure and function are complementary properties: one tells you nothing about the other. If you have no idea what the function of a neuron is, you are not going to figure it out by looking at one. The reverse is also true. If you know what the function of a neuron is, you still would have no idea of its appearance. Without any prior knowledge, the function of the neurons can't be derived from their structure, nor can their structure be derived from their function. They are two separate, irreducible layers with different protocols.

The enterprise of learning more about the underlying parts of the brain needs to expand its agenda and also focus on neural design. Simply trying to locate the structure that produces consciousness, as Descartes and many of his predecessors have attempted, will not unveil the Holy Grail, because consciousness is inherent *throughout* the brain. Cutting huge chunks from the cortex does not disrupt consciousness, but only changes its contents. It is not compartmentalized in the brain like many other mental capacities, such as speech production or visual processing, but is a crucial element of all these various capacities. Again, as I have discussed, the most compelling evidence for piecemeal consciousness is revealed through the minds of split-brain patients: When transmission between the hemispheres is severed, each will continue to have its own conscious experience.

While it is not intuitive to think that our consciousness emanates from several independent sources, this appears to be the brain's design. Once this concept is fully grasped, the true challenge will be to understand how the design principles of the brain allow for consciousness to emerge in this manner. This is the future challenge for brain science.

A Final Word

When I started this book, I didn't think I would wind up with some of the thoughts I have now outlined. The lurking question was always: Is consciousness really an instinct?

In his now classic book *The Language Instinct*, Steven Pinker provides a necessary wake-up call for the scientific community: How can minds and brains be both delivered biologically and also modified by experience? The book provided a needed framework for thinking about the limits of learning and the realities of mind parts derived through natural selection. Pinker also brilliantly observed that conceptualizing higher-order human traits (such as language) as *instincts* is downright jarring.

Plopping the phenomenon of consciousness onto the instinct list—right in there with anger, shyness, affection, jealousy, envy, rivalry, sociability, and so on—is equally disorienting. Instincts, as we all know, evolve gradually, making us more fit for our environment. Adding consciousness to the instinct list suggests that this precious human property, which we all hold dear, is not a miraculously endowed part of our species' special hardware.

If we allow consciousness to be an instinct, we toss it into the vast biological world with all of its history, richness, variation, and continuum. Where did it come from? How did it evolve? What other species share features of it?

Let's pause to ask the fundamental question: What is an instinct, anyway? The term is thrown around like confetti at a parade. Each year, the list of instincts grows and grows. You would almost think that if you popped off the skull, you would see a bunch of labeled lines, each representing one of the much heralded instincts. Indeed, the human brain *ought* to be a rat's nest of wires connected up to do their job. Yet if you ask a neuroscientist to show you the network for a particular instinct, such as rivalry or sociability, no such knowledge exists—at least, not yet. So how does it help to call stuff instincts?

When feeling at sea about definitions and meanings in the mind/brain business, it is always rewarding to dial up William James once again. More than 125 years ago, James wrote a landmark article simply titled "What Is an Instinct?" He wastes no time in defining the concept:

> Instinct is usually defined as the faculty of acting in such a way as
> to produce certain ends, without foresight of the ends, and without
> previous education in the performance. . . . [Instincts] are the
> functional correlatives of structure. With the presence of a certain
> organ goes, one may say, almost always a native aptitude for its
> use. "Has the bird a gland for the secretion of oil? She knows
> instinctively how to press the oil from the gland, and apply it to
> the feather."[5]

The definition seems straightforward, and yet it is cleverly dualistic. An instinct calls upon a physical structure to function. Yet using the structure calls upon an "aptitude," which apparently comes along for free. Finding the physical correlates of an instinct's physical apparatus is doable, but how do we learn how it comes to be used? Does it just happen? Not a very scientific answer. Does the bird start out with a reflex to press the gland and, over time, learn that, as a consequence, everything works better? Clearly if there was no oil gland, there would be no oil and no opportunity for learning to use it to fly better. One can see the blind loop of natural selection and experience working together to form what we would call an instinct.

Bird behavior is one thing, but does this really apply to human cognition and consciousness? James offers a rationale for how it might all work:

> A single complex instinctive action may involve successively the awakening of impulses. . . . Thus a hungry lion starts to *seek* prey by the awakening in him of imagination coupled with desire; he begins to *stalk* it when, on eye, ear, or nostril, he gets an impression of its presence at a certain distance; he *springs* upon it, either when the booty takes alarm and flees, or when the distance is sufficiently reduced; he proceeds to *tear* and *devour* it the moment he gets a sensation of its contact with his claws and fangs. Seeking, stalking, springing, and devouring are just so many different kinds of muscular contraction, and neither kind is called forth by the stimulus appropriate to the other.[6]

As I look at James's work now, I recognize a schema that fits the module/layering ideas. James appears to suggest that the structural aspects of instincts are inbuilt modules embedded in a layered architecture. Each instinct can function independently for simple behaviors, but they also work as a confederation. Individual instincts can be sequenced in a coordinated fashion for more complex actions that make them look an awful lot like higher-order instincts. The avalanche of sequences is what we call consciousness. James argues that the *competitive* dynamics that go into the sequencing of basic instincts can produce what appears to be a more complex behavior manifested from a complex internal state. He even adds a description of the animal's experience of obeying an instinct: "Every impulse and every step of every instinct shines with its own sufficient light, and seems at the moment the only eternally right and proper thing to do. It is done for its own sake exclusively." It sounds like a lot of bubbles are conjoined by the arrow of time and produce something like what we call conscious experience.

The dynamics of which bubble pops up when is no doubt influenced by experience and learning. However, experience, learning, and consciousness must all be isomorphic—operational within the same system. Once the phenomenon is thought of in this way, we see conscious experience for what it is: Mother Nature's trick. Thinking of consciousness as an evolved instinct (or a whole sequence of them) shows us where to look for how it emerged from the cold inanimate world. It opens our eyes to the realization that each aspect of a conscious experience is the unfolding of other instincts that humans possess, and that, by their very nature, the mechanisms and capacities they harbor produce the felt state of conscious experience. Remarkably, in the past few years biologists of all stripes have been able to come together in a breathtaking way to identify twenty-nine specific networks in the brain of a fly, each controlling a specific behavior. These individual behaviors can be flexibly combined and recombined into more complex patterns. Yes, it is in the fruit fly where we may learn the lessons of consciousness! The hunt for understanding the physical dimension of instincts is on.[7]

However, many abhor the use of concepts such as instinct to describe phenomenal conscious experience. To them, this definition also robs humans of their unique status in the animal kingdom, namely, that we alone are morally responsible for our actions. Humans can choose to do, and we can therefore choose to "do the right thing." If consciousness is an instinct, they argue, then humans must be automatons, or witless zombies. Yet, putting aside for the moment the physical realities of quantum mechanics and *Schnitts* with their liberating symbolic functioning, we can argue that accepting the idea that a complex entity like the brain/body/mind has a knowable mechanism does not doom one to such deterministic and despairing views. James himself addressed this overarching concern:

Here we immediately reap the good fruits of our simple physiological conception of what an instinct is. If it be a mere excito-motor impulse, due to the pre-existence of a certain "reflex-arc" in the nerve-centres of the creature, of course it must follow the law of all such reflex-arcs. One liability of such arcs is to have their activity "inhibited" by other processes going on at the same time. It makes no difference whether the arc be organized at birth, or ripen spontaneously later, or be due to acquired habit, it must take its chances with all the other arcs, and sometimes succeed, and sometimes fail. . . . The mystical view of an instinct would make it invariable. The physiological view would require it to show occasional irregularities in any animal in whom the number of separate instincts, and the possible entrance of the same stimulus into several of them, were great. And such irregularities are what every superior animal's instincts do show in abundance.[8]

James provides much more, and it does take time to absorb the idea of instincts. I urge you to read his original paper to see his clear thinking, clear writing, and unshakable pragmatism on these difficult issues. James points the way forward, refusing to accept the despairing caricature of humankind as robot at the beck and call of reflex responses. To him, a complex behavioral state can be produced by varying the combinations of simple, independent modules, just as a combination of multiple different small movements makes the complex behavior of a pole vaulter as he sails upward over the pole. When acting together in a coordinated way, even simple systems can make observers believe other forces exist. James's stance is clearly stated: "My first act of free will shall be to believe in free will." This proclamation is consistent with the idea that beliefs, ideas, and thoughts can be part of the mental system. The symbolic representations within this system, with all their flexibility and arbitrariness, are very much tied to the physical mechanisms of the brain. Ideas do have consequences, even in the physically constrained brain. No despair called for: mental states can influence physical action in the top-down way!

The flexibility of my own symbolic representations has been a source of joy and surprise, not despair, over the course of this project. Perhaps the most surprising discovery for me is that I now think we humans will never build a machine that mimics our personal consciousness. Inanimate silicon-based machines work one way, and living carbon-based systems work another. One works with a deterministic set of instructions, and the other through symbols that inherently carry some degree of uncertainty.

This perspective leads to the view that the human attempt to mimic intelligence and consciousness in machines, a continuing goal in the field of AI, is doomed. If living systems work on the principle of complementarity—the idea that the physical side is mirrored with an arbitrary symbolic side, with symbols that are the result of natural selection—then purely deterministic models of what makes life will always fall short. In an AI model, the memory for an event is in one place and can be deleted with one keystroke. In a living, layered symbolic system, however, each aspect of a mechanism can be switched out for another symbol, so long as each plays its proper role. It is this way because it is what life itself allows, indeed demands: complementarity.

Who is going to put science to all of these ideas? What will the neuroscience of tomorrow look like? In my opinion, the hunt for enduring answers will have to include neuroengineers, with their ability to eke out the deep principles of the design of things. Such a revolution is in its early days, but the perspective it offers is clear. A layered architecture, which allows the option of adding supplemental layers, offers a framework to explain how brains became increasingly complex through the process of natural selection while conserving successful basic features. One challenge is to identify what the various processing layers do, and the bigger challenge is to crack the protocols that allow one layer to interpret the processing results of its neighbor layers. That will involve crossing the *Schnitt*, that epistemic gap that links subjective experience with objective processing, which has been around since the first living cell. Capturing how the physical side of the gap, the neurons, works with the symbolic side, the mental dimensions, will be achieved through the language of complementarity.

In the end, we must realize that consciousness is an instinct. Consciousness is part of organismic life. We never have to learn how to produce it or how to utilize it. On a recent trip to Charleston, my wife and I were out in the countryside looking for some good ole fried chicken and cornbread. We finally found a small roadside diner and ordered. As the waitress was walking away, I said, "Oh yes, and add some grits to that order." She turned back to me, smiled, and said, "Honey, grits come." Grits come with the order, and so does what we call consciousness. We are lucky for both.

NOTES

1. History's Rigid, Rocky, and Goofy Way of Thinking About Consciousness

1. Zan Boag, "Searle: It upsets me when I read the nonsense written by my contemporaries," *NewPhilosopher* 2, January 25, 2014, http://www.newphilosopher.com/articles/john-searle-it-upsets-me-when-i-read-the-nonsense-written-by-my-contemporaries/.
2. Henri Frankfort et al., *The Intellectual Adventure of Ancient Man: An Essay of Speculative Thought in the Ancient Near East* (Chicago: University of Chicago Press, 1977).
3. Robert Rosen, *Life Itself: A Comprehensive Inquiry into the Nature, Origin, and Fabrication of Life* (New York: Columbia University Press, 1991), 20.
4. René Descartes, *Discourse on Method* (1637), in Robert Hutchins, Mortimer J.

Adler, and Wallace Brockway, eds., *Great Books of the Western World*, vol. 31, *Descartes/Spinoza* (Chicago: Encyclopaedia Britannica, 1952), 51.

5. Gary Hatfield, "René Descartes," in *Stanford Encyclopedia of Philosophy Archive*, 2014, http://plato.stanford.edu/archives/fall2015/entries/descartes/.

6. René Descartes, *The Philosophical Writings of Descartes*, vol. 3, *The Correspondence*, ed. and trans. John Cottingham, Robert Stoothoff, Dugald Murdoch, and Anthony Kenny (Cambridge, U.K.: Cambridge University Press, 1984), 19-20.

2. The Dawn of Empirical Thinking in Philosophy

1. John Locke, *An Essay Concerning Human Understanding*, in Hutchins, Adler, and Brockway, *Great Books of the Western World*, vol. 35, *Locke/Berkeley/Hume*, 2.1.19.

2. David Hume, "A Letter to a Physician" (1734), in *Life and Correspondence of David Hume*, ed. John Hill Burton (Edinburgh: William Tait, 1846), 35.

3. Robert G. Brown, "Philosophy Is Bullshit: David Hume," in *Axioms as the Basis for All Understanding*, 2003, retrieved February 10, 2016, from https://www.phy.duke.edu/~rgb/Beowulf/axioms/axioms/node4.html.

4. David Hume, *A Treatise of Human Nature: Being an Attempt to Introduce the Experimental Method of Reasoning into Moral Subjects*, vol. 1, *Of the Understanding* (London: John Noon, 1739), T intro.4, SBN xv, http://www.davidhume.org/texts/thn.html.

5. Ibid., T 1.1.1.7, SBN 4.

6. David Hume, *An Abstract of A Book Lately Published; Entituled, A Treatise of Human Nature, &c.* (London: C. Borbet, 1740), SBN 662, http://www.davidhume.org/texts/abs.html.

7. David Hume, *An Enquiry Concerning Human Understanding* (1748), in Hutchins, Adler, and Brockway, eds., *Great Books of the Western World*, vol. 35, *Locke/Berkeley/Hume*, 458.

8. David Hume, to John Stewart (1754), letter 91 in *The Letters of David Hume*, vol. 1, *1727–1765*, ed. J. Y. T. Greig (1932; repr. Oxford and New York: Oxford University Press, 2011), 187.

9. Hume, *Treatise of Human Nature*, T 1.4.6.3.

10. Ibid., T 1.4.6.6, SBN 254.

11. Ibid., T 1.4.6.4, SBN 253.

12. Arthur Schopenhauer, *Essays and Aphorisms*, trans. R. J. Hollingdale (1851; repr. London: Penguin Group, 2004), 223.

13. Arthur Schopenhauer, *The World as Will and Representation* (1818), trans. E. F. J. Payne (1958; repr. New York: Dover, 1996), 2:209, https://digitalseance .files.wordpress.com/2010/07/32288614-schopenhauer-the-world-as-will-and -representation-v2.pdf.

14. Arthur Schopenhauer, *The World as Will and Idea* (1818), trans. R. B. Haldane and J. Kemp (London: Routledge and Kegan Paul Ltd., 1883), 3:127.

15. Cubie King and David Von Drehle, "Encounters with the Arch-Genius, David Gelernter," *Time*, February 25, 2016, http://time.com/4236974/encounters -with-the-archgenius/.

16. Schopenhauer, *World as Will and Representation*, 2:136.

17. Hermann von Helmholtz, *Treatise on Physiological Optics* (1867), ed. James P. C. Southall (1924; repr. New York: Dover, 1962, 2005), vol. 3.

18. Henry Maudsley, *The Physiology and Pathology of Mind* (New York: D. Apple-ton and Company, 1867), 15, https://archive.org/stream/physiologypathol00 maudiala#page/14/mode/2up/search/unconscious+mental+activity.

19. Ibid., 120.

20. Francis Galton, "Psychometric Experiments," *Brain* 2 (1879), 149–62.

21. Owen Flanagan, *The Science of the Mind* (Cambridge, Mass.: MIT Press, 1984), 60.

22. Franz Brentano, *Psychology from an Empirical Standpoint* (1874), ed. Oskar Kraus, (Eng.) Linda L. McAlister, trans. Antos C. Rancurello, D. B. Terrell, and Linda L. McAlister, International Library of Philosophy (London and New York: Routledge, 1995), 68, http://14.139.206.50:8080/jspui/bitstream /1/1432/1/Brentano,%20Franz%20-%20Psychology%20from%20an%20Em pirical%20Standpoint.pdf.

23. Flanagan, *Science of the Mind*, 62.

24. Drew Westen, "The Scientific Legacy of Sigmund Freud: Toward a Psychody-namically Informed Psychological Science," *Psychological Bulletin* 124 (1998), 333.

25. Charles Darwin, *On the Origin of Species by Means of Natural Selection, or the Preservation of Favoured Races in the Struggle for Life*, first American edi-tion, fourth printing, revised and augmented (New York: D. Appleton, 1860), 424.

26. Darwin, *The Descent of Man and Selection in Relation to Sex*, in Hutchins, Adler, and Brockway, *Great Books of the Western World*, vol. 49, *Darwin*, 319.

27. Darwin, *Origin of Species*, 425.

3. Twentieth-Century Strides and Openings to Modern Thought

1. William James, *Pragmatism: A New Name for Some Old Ways of Thinking*, Lecture 1 (New York: Longmans, Green, and Co., 1907), 6–7, https://archive .org/stream/157unkngoog#page/n26/mode/2up/search/clash+of+human +temperaments.
2. Ibid., 13–14.
3. Ibid., 15.
4. Ibid., 65–66.
5. Michael I. Posner, *Chronometric Explorations of Mind* (Hillsdale, N.J.: Lawrence Erlbaum Associates, 1978).
6. Wilder Penfield, "Speech, Perception and the Uncommitted Cortex," in John C. Eccles, ed., *Brain and Conscious Experience* (New York: Springer-Verlag, 1966), 234.
7. Ibid., 235.
8. George A. Miller, *Psychology: The Science of Mental Life* (New York: Harper and Row, 1962), 25.
9. David R. Curtis and Per Andersen, "Sir John Carew Eccles, A.C. 27 January 1903—2 May 1997," *Biographical Memoirs of Fellows of the Royal Society* 47 (2001), 160–87, https://www.science.org.au/fellowship/fellows/biographical -memoirs/john-carew-eccles-1903-1997#2.
10. Karl R. Popper and John C. Eccles, *The Self and Its Brain: An Argument for Interactionism* (Berlin: Springer-Verlag, 1977).
11. Henry H. Dale, "The Beginnings and the Prospects of Neurohumoral Transmission," *Pharmacological Reviews* 6 (1954), 7–13.
12. John C. Eccles, "Hypotheses Relating to the Brain-Mind Problem," *Nature* 168 (1951), 53–57.
13. E. G. Walsh, "[Review of] *Brain and Conscious Experience: Study Week September 28 to October 4, 1964 of the Pontificia Academia Scientiarum*. Edited by Sir John C. Eccles. Berlin, Heidelberg, New York: Springer-Verlag . . . ," *Quarterly Journal of Experimental Physiology and Cognate Medical Sciences* 52 (1967), 330.
14. William H. Thorpe, "Ethology and Consciousness," in Eccles, *Brain and Conscious Experience*, 44.
15. John C. Eccles, "Conscious Experience and Memory," in Eccles, *Brain and Conscious Experience*, 326.

16. Roger W. Sperry, "Brain Bisection and Mechanisms of Consciousness," in Eccles, *Brain and Conscious Experience*, 299.

17. Roger W. Sperry, "Mind-Brain Interaction: Mentalism, Yes; Dualism, No," *Neuroscience* 5 (1980), 196.

18. Sperry, "Brain Bisection," 308.

19. Ibid.

20. Eccles, *Brain and Conscious Experience*, 250.

21. Ibid., 248.

22. Charles G. Gross, "Hans-Lukas Teuber: A Tribute," *Cerebral Cortex* 4 (1994), 451–54.

23. Eccles, *Brain and Conscious Experience*, 582.

24. Roger W. Sperry, "Mind, Brain, and Humanist Values," *Bulletin of the Atomic Scientists* 22 (1966), 2–6.

25. Roger W. Sperry, "Perception in the Absence of the Neocortical Commissures," in David A. Hamburg, Karl H. Pribram, and Albert J. Stunkard, eds., *Perception and Its Disorders*, vol. 48 (Baltimore: Williams and Wilkins, 1970), 123–28.

26. Donald M. MacKay, "Soul, Brain Science and the" entry in R. L. Gregory, ed., *The Oxford Companion to the Mind* (Oxford: Oxford University Press, 1987), 724–25.

27. P. M. S. Hacker, "The Sad and Sorry History of Consciousness: Being, among Other Things, a Challenge to the 'Consciousness-Studies Community,'" *Royal Institute of Philosophy Supplement* 70 (2012), 149–68.

28. Thomas Nagel, "The Psychophysical Nexus," in Paul Boghossian and Christopher Peacocke, eds., *New Essays on the A Priori* (Oxford: Oxford University Press, 2000), 432–71.

29. Douglas R. Hofstadter and Daniel C. Dennett, *The Mind's I: Fantasies and Reflections on Self and Soul* (New York: Basic Books, 1981, 2000), 409.

30. Owen Flanagan, *The Problem of the Soul: Two Visions of Mind and How to Reconcile Them* (New York: Basic Books, 2002), 87.

31. Francis H. Crick, "Thinking About the Brain," *Scientific American* 241 (1979), 219–32.

32. Michael I. Posner and Mary K. Rothbart, "Attentional Mechanisms and Conscious Experience," in A. D. Milner and M. D. Rugg, eds., *The Neuropsychology of Conscious Experience* (London: Academic Press, 1992), 97–117.

33. Michael S. Gazzaniga, *The Bisected Brain* (New York: Appleton Century Crofts, 1970).

34. Crick, "Thinking About the Brain."
35. Ibid.
36. Ibid.
37. Francis Crick and Christof Koch, "Towards a Neurobiological Theory of Consciousness," *Seminars in the Neurosciences* 2 (1990), 263–75.
38. Christof Koch, *The Quest for Consciousness: A Neurobiological Approach* (Englewood, Colo.: Roberts and Company, 2004), 17.
39. Ibid., 15.

4. Making Brains One Module at a Time

1. Charles S. Sherrington, *Man on His Nature: The Gifford Lectures, 1937–38* (1940; repr. Cambridge, U.K.: Cambridge University Press, 2009).
2. Michael S. Gazzaniga, "Brain Mechanisms and Conscious Experience," *Experimental and Theoretical Studies of Consciousness*, CIBA Foundation Symposium 174 (Chichester, U.K.: John Wiley and Sons, 1993), 247–62.
3. Edoardo Bisiach and Claudio Luzzatti, "Unilateral Neglect of Representational Space," *Cortex* 14 (1978), 129–33.
4. Patrik Vuilleumier, "Mapping the Functional Neuroanatomy of Spatial Neglect and Human Parietal Lobe Functions: Progress and Challenges," *Annals of the New York Academy of Sciences* 1296 (2013), 50–74.
5. Bruce T. Volpe, Joseph E. Ledoux, and Michael Gazzaniga, "Information Processing of Visual Stimuli in an 'Extinguished' Field," *Nature* 282 (1979), 722–24.
6. Reinhold Messner, *The Naked Mountain* (Seattle: The Mountaineers Books, 2003), 299.
7. W. Dewi Rees, "The Hallucinations of Widowhood," *British Medical Journal* 4 (1971), 37.
8. Shahar Arzy et al., "Induction of an Illusory Shadow Person," *Nature* 443 (2006), 287.
9. Olaf Blanke et al., "Neurological and Robot-Controlled Induction of an Apparition," *Current Biology* 24 (2014), 2681–86.
10. Ibid.
11. Frederico A. C. Azevedo et al., "Equal Numbers of Neuronal and Nonneuronal Cells Make the Human Brain an Isometrically Scaled-up Primate Brain," *Journal of Comparative Neurology* 513 (2009), 532–41.
12. Suzana Herculano-Houzel, "The Human Brain in Numbers: A Linearly Scaled-up Primate Brain," *Frontiers in Human Neuroscience* 3 (2009), 31.

13. Mark E. Nelson and James M. Bower, "Brain Maps and Parallel Computers," *Trends in Neurosciences* 13 (1990), 403–8.

14. Donald D. Clarke and Louis Sokoloff, "Circulation and Energy Metabolism of the Brain," in George J. Siegel et al., eds., *Basic Neurochemistry: Molecular, Cellular and Medical Aspects*, 6th ed. (Philadelphia: Lippincott-Raven, 1999), 637–70.

15. Georg F. Striedter, *Principles of Brain Evolution* (Sunderland, Mass.: Sinauer Associates, 2005).

16. David Meunier, Renaud Lambiotte, and Edward T. Bullmore, "Modular and Hierarchically Modular Organization of Brain Networks," *Frontiers in Neuroscience* 4 (2010), 200.

17. Ibid.

18. Dmitri B. Chklovskii, Thomas Schikorski, and Charles F. Stevens, "Wiring Optimization in Cortical Circuits," *Neuron* 34 (2002), 341–47.

19. Danielle S. Bassett et al., "Dynamic Reconfiguration of Human Brain Networks during Learning," *Proceedings of the National Academy of Sciences* (*PNAS*) 108 (2011), 7641–46; Danielle S. Bassett et al., "Robust Detection of Dynamic Community Structure in Networks," *Chaos: An Interdisciplinary Journal of Nonlinear Science* 23 (2013), 013142.

20. Olaf Sporns and Richard F. Betzel, "Modular Brain Networks," *Annual Review of Psychology* 67 (2016), 613–40.

21. Striedter, *Principles of Brain Evolution*, 248.

22. Beth L. Chen, David H. Hall, and Dmitri B. Chklovskii, "Wiring Optimization Can Relate Neuronal Structure and Function," *PNAS* 103 (2006), 4723–28; Christopher Cherniak et al., "Global Optimization of Cerebral Cortex Layout," *PNAS* 101 (2004), 1081–86; Yong-Yeol Ahn, Hawoong Jeong, and Beom Jun Kim, "Wiring Cost in the Organization of a Biological Neuronal Network," *Physica A: Statistical Mechanics and Its Applications* 367 (2006), 531–37.

23. Jeff Clune, Jean-Baptiste Mouret, and Hod Lipson, "The Evolutionary Origins of Modularity," *Proceedings of the Royal Society of London B: Biological Sciences* 280 (2013), 20122863.

24. Peter Carruthers, *The Architecture of the Mind: Massive Modularity and the Flexibility of Thought* (Oxford: Oxford University Press, 2006).

25. Sporns and Betzel, "Modular Brain Networks."

26. Nicola Clayton and Nathan Emery, "Corvid Cognition," *Current Biology* 15 (2005), R80–R81.

27. Alex H. Taylor et al., "Complex Cognition and Behavioural Innovation in New Caledonian Crows," *Proceedings of the Royal Society of London B: Biological Sciences* 277 (2010), 2637–43.

28. Jennifer C. Holzhaider, Gavin R. Hunt, and Russell D. Gray, "Social Learning in New Caledonian Crows," *Learning and Behavior* 38 (2010), 206–19.

29. Gavin R. Hunt, C. Lambert, and Russell D. Gray, "Cognitive Requirements for Tool Use by New Caledonian Crows (*Corvus moneduloides*)," *New Zealand Journal of Zoology* 34 (2007), 1–7.

30. Andrew Whiten et al., "Emulation, Imitation, Over-Imitation and the Scope of Culture for Child and Chimpanzee," *Philosophical Transactions of the Royal Society of London B: Biological Sciences* 364 (2009), 2417–28.

31. Wolfgang Köhler, trans. Ella Winter, *The Mentality of Apes* (London: Kegan Paul, Trench, Trübner and Company, 1925).

32. Kristin Liebal et al., "Infants Use Shared Experience to Interpret Pointing Gestures," *Developmental Science* 12 (2009), 264–71.

33. David Premack, "Why Humans Are Unique: Three Theories," *Perspectives on Psychological Science* 5 (2010), 22–32.

34. Carruthers, *Architecture of the Mind*.

35. David Premack and Guy Woodruff, "Does the Chimpanzee Have a Theory of Mind?" *Behavioral and Brain Sciences* 1 (1978), 515–26.

36. Josep Call and Michael Tomasello, "Does the Chimpanzee Have a Theory of Mind? 30 Years Later," *Trends in Cognitive Sciences* 12 (2008), 187–92.

37. Zijing He, Matthias Bolz, and Renée Baillargeon, "Understanding of False Belief in 2.5-year-olds in a Violation-of-Expectation Test," paper presented at the Biennial Meeting of the Society for Research in Child Development, Boston, March 2007.

38. Christopher Krupenye et al., "Great Apes Anticipate That Other Individuals Will Act According to False Beliefs," *Science* 354 (2016), 110–14.

39. John W. Pilley and Alliston K. Reid, "Border Collie Comprehends Object Names as Verbal Referents," *Behavioural Processes* 86 (2011), 184–95; John W. Pilley, "Border Collie Comprehends Sentences Containing a Prepositional Object, Verb, and Direct Object," *Learning and Motivation* 44 (2013), 229–40.

40. Katharina C. Kirchhofer et al., "Dogs (*Canis familiaris*), but Not Chimpanzees (*Pan troglodytes*), Understand Imperative Pointing," *PloS One* 7 (2012), e30913.

41. Michelle E. Maginnity and Randolph C. Grace, "Visual Perspective Taking by Dogs (*Canis familiaris*) in a Guesser–Knower Task: Evidence for a Canine Theory of Mind?" *Animal Cognition* 17 (2014), 1375–92.

42. Brian Hare and Michael Tomasello, "Human-like Social Skills in Dogs?" *Trends in Cognitive Sciences* 9 (2005), 439–44.

43. Muhammad A. Spocter et al., "Neuropil Distribution in the Cerebral Cortex Differs between Humans and Chimpanzees," *Journal of Comparative Neurology* 520 (2012), 2917–29.

44. Julia Mehlhorn et al., "Tool-Making New Caledonian Crows Have Large Associative Brain Areas," *Brain, Behavior and Evolution* 75 (2010), 63–70.

45. Justin S. Feinstein et al., "The Human Amygdala and the Induction and Experience of Fear," *Current Biology* 21 (2011), 34–38.

5. The Beginnings of Understanding Brain Architecture

1. Robert Rosen, *Dynamical System Theory in Biology* (New York: Wiley, 1970).

2. Michael Polanyi, "Life's Irreducible Structure," *Science* 160 (1968), 1308.

3. Ibid.

4. Marie E. Csete and John C. Doyle, "Reverse Engineering of Biological Complexity," *Science* 295 (2002), 1664–69.

5. John C. Doyle and Marie E. Csete, "Architecture, Constraints, and Behavior," *PNAS* 108, Supplement 3 (2011), 15624–30.

6. Polanyi, "Life's Irreducible Structure."

7. Doyle and Csete, "Architecture, Constraints, and Behavior."

8. David L. Alderson and John C. Doyle, "Contrasting Views of Complexity and Their Implications for Network-Centric Infrastructures," *IEEE Transactions on Systems, Man and Cybernetics—Part A: Systems and Humans* 40 (2010), 840.

9. Ibid.

10. Doyle and Csete, "Architecture, Constraints, and Behavior."

11. Jerzy Wegiel et al., "The Neuropathology of Autism: Defects of Neurogenesis and Neuronal Migration, and Dysplastic Changes," *Acta Neuropathologica* 119 (2010), 755–70.

12. Aswin Sekar et al., "Schizophrenia Risk from Complex Variation of Complement Component 4," *Nature* 530 (2016), 177–83.

13. Alderson and Doyle, "Contrasting Views of Complexity."

14. Mung Chiang et al., "Layering as Optimization Decomposition: A Mathematical Theory of Network Architectures," *Proceedings of the IEEE* 95 (2007), 255–312.

15. Harold Pashler, *Encyclopedia of the Mind*, vol. 1 (Thousand Oaks, Calif.: Sage Publications, 2013), 465.

16. Tony J. Prescott, Peter Redgrave, and Kevin Gurney, "Layered Control Architectures in Robots and Vertebrates," *Adaptive Behavior* 7 (1999), 99–127.

17. Ibid., 101.

18. Marc Kirschner and John Gerhart, "Evolvability," *PNAS* 95 (1998), 8420–27.

19. Peter T. Boag and Peter R. Grant, "Intense Natural Selection in a Population of Darwin's Finches (Geospizinae) in the Galápagos," *Science* 214 (1981), 82–85.

20. John Gerhart and Marc Kirschner, "The Theory of Facilitated Variation," *PNAS* 104, supplement 1 (2007), 8582–89.

21. Alderson and Doyle, "Contrasting Views of Complexity."

22. Doyle and Csete, "Architecture, Constraints, and Behavior."

23. Ibid.

24. Christopher W. Johnson, "What Are Emergent Properties and How Do They Affect the Engineering of Complex Systems?" *Reliability Engineering and System Safety* 91 (2006), 1475–81.

25. Eve Marder, "Variability, Compensation and Modulation in Neurons and Circuits," *PNAS* 108, supplement 3 (2011), 15542–48.

26. Tamar Friedlander et al., "Evolution of Bow-Tie Architectures in Biology," *PLoS Computational Biology* 11 (2015), e1004055.

27. John C. Doyle, "Guaranteed Margins for LQG Regulators," *IEEE Transactions on Automatic Control* 23 (1978), 756–57.

28. Alderson and Doyle, "Contrasting Views of Complexity."

29. Arne J. Nagengast, Daniel A. Braun, and Daniel M. Wolpert, "Risk-Sensitive Optimal Feedback Control Accounts for Sensorimotor Behavior Under Uncertainty," *PLoS Computational Biology* 6 (2010), e1000857.

30. Fiona A. Chandra, Gentian Buzi, and John C. Doyle, "Glycolytic Oscillations and Limits on Robust Efficiency," *Science* 333 (2011), 187–92.

31. Daniel Kahneman, *Thinking, Fast and Slow* (New York: Farrar, Straus and Giroux, 2011).

32. Roger W. Sperry, "Neurology and the Mind-Brain Problem," *American Scientist* 40 (1952), 291–312.

33. Hear him do it at https://www.ted.com/talks/daniel_wolpert_the_real_reason_for_brains?language=en.

6. Gramps Is Demented but Conscious

1. David A. Drachman, "The Amyloid Hypothesis, Time to Move On: Amyloid Is the Downstream Result, Not Cause, of Alzheimer's Disease," *Alzheimer's and Dementia* 10 (2014), 372–80; Jessica Freiherr et al., "Intranasal Insulin as

a Treatment for Alzheimer's Disease: A Review of Basic Research and Clinical Evidence," *CNS Drugs* 27 (2013), 505–14.

2. Stanley B. Klein, Leda Cosmides, and Kristi A. Costabile, "Preserved Knowledge of Self in a Case of Alzheimer's Dementia," *Social Cognition* 21 (2003), 157–65.

3. Lydia Krabbendam and Jim van Os, "Schizophrenia and Urbanicity: A Major Environmental Influence—Conditional on Genetic Risk," *Schizophrenia Bulletin* 31 (2005), 795–99.

4. Elizabeth Cantor-Graae and Jean-Paul Selten, "Schizophrenia and Migration: A Meta-Analysis and Review," *American Journal of Psychiatry* 162 (2005), 12–24.

5. Wim Veling et al., "Ethnic Density of Neighborhoods and Incidence of Psychotic Disorders among Immigrants," *American Journal of Psychiatry* 165 (2008), 66–73.

6. Theresa H. M. Moore et al., "Cannabis Use and Risk of Psychotic or Affective Mental Health Outcomes: A Systematic Review," *The Lancet* 370 (2007), 319–28; Robin M. Murray et al., "Cannabis, the Mind and Society: The Hash Realities," *Nature Reviews Neuroscience* 8 (2007), 885–95.

7. Kurt Schneider, *Clinical Psychopathology*, trans. M. W. Hamilton (New York: Grune and Stratton, 1959).

8. Jim van Os and Shitij Kapur, "Schizophrenia," *The Lancet* 374 (2009), 635–45; Jim van Os, " 'Salience Syndrome' Replaces 'Schizophrenia' in DSM-V and ICD-11: Psychiatry's Evidence-Based Entry into the 21st Century?" *Acta Psychiatrica Scandinavica* 120 (2009), 363–72.

9. Shitij Kapur, "Psychosis as a State of Aberrant Salience: A Framework Linking Biology, Phenomenology, and Pharmacology in Schizophrenia," *American Journal of Psychiatry* 160 (2003), 13–23.

10. Marc Laruelle, "Imaging Dopamine Transmission in Schizophrenia: A Review and Meta-Analysis," *Quarterly Journal of Nuclear Medicine* 42 (1998), 211–21; Oliver Guillin, Anissa Abi-Dargham, and Marc Laruelle, "Neurobiology of Dopamine in Schizophrenia," *International Review of Neurobiology* 78 (2007), 1–39.

11. Kapur, "Psychosis as a State of Aberrant Salience."

12. Jimmy Jensen et al., "The Formation of Abnormal Associations in Schizophrenia: Neural and Behavioral Evidence," *Neuropsychopharmacology* 33 (2008), 473–79; J. P. Roiser et al., "Do Patients with Schizophrenia Exhibit Aberrant Salience?" *Psychological Medicine* 39 (2009), 199–209.

13. Rosalind Cartwright, "Sleepwalking Violence: A Sleep Disorder, a Legal Dilemma, and a Psychological Challenge," *American Journal of Psychiatry* 161 (2004), 1149–58.

14. Pierre Maquet et al., "Functional Neuroanatomy of Human Slow Wave Sleep," *Journal of Neuroscience* 17 (1997), 2807–12; A. R. Braun et al., "Regional Cerebral Blood Flow throughout the Sleep-Wake Cycle: An $H_2{}^{15}O$ PET Study," *Brain* 120 (1997), 1173–97; Jesper L. R. Andersson et al., "Brain Networks Affected by Synchronized Sleep Visualized by Positron Emission Tomography," *Journal of Cerebral Blood Flow and Metabolism* 18 (1998), 701–15; C. Kaufmann et al., "Brain Activation and Hypothalamic Functional Connectivity During Human Non–Rapid Eye Movement Sleep: An EEG/fMRI Study," *Brain* 129 (2006), 655–67.

15. Claudio Bassetti et al., "SPECT During Sleepwalking," *The Lancet* 356 (2000), 484–85.

16. Michele Terzaghi et al., "Evidence of Dissociated Arousal States during NREM Parasomnia from an Intracerebral Neurophysiological Study," *Sleep* 32 (2009), 409–12.

17. Steven Laureys et al., "The Locked-in Syndrome: What Is It Like to Be Conscious but Paralyzed and Voiceless?" *Progress in Brain Research* 150 (2005), 495–511.

18. Jean-Dominique Bauby, *The Diving Bell and the Butterfly: A Memoir of Life in Death* (New York: Alfred A. Knopf, 1997).

19. Ibid.

20. Sofiane Ghorbel, "Statut fonctionnel et qualité de vie chez le locked-in syndrome a domicile," *DEA Motricité Humaine et Handicap* (Saint-Etienne, Montpellier, France: Laboratory of Biostatistics, Epidemiology and Clinical Research, Université Jean Monnet, 2002).

21. Ronald E. Cranford, "The Persistent Vegetative State: The Medical Reality (Getting the Facts Straight)," *Hastings Center Report* 18 (1988), 27–32.

22. Steven Laureys, Olivia Gosseries, and Giulio Tononi, eds., *The Neurology of Consciousness: Cognitive Neuroscience and Neuropathology*, 2nd ed. (Amsterdam: Academic Press, 2016).

23. Björn Merker, "Consciousness Without a Cerebral Cortex: A Challenge for Neuroscience and Medicine," *Behavioral and Brain Sciences* 30 (2007), 63–81.

24. Jaak Panksepp et al., "Effects of Neonatal Decortication on the Social Play of Juvenile Rats," *Physiology and Behavior* 56 (1994), 429–43.

25. Jaak Panksepp, "Affective Consciousness: Core Emotional Feelings in Animals and Humans," *Consciousness and Cognition* 14 (2005), 30–80.

26. David J. Anderson and Ralph Adolphs, "A Framework for Studying Emotions across Species," *Cell* 157 (2014), 187–200.

27. Jaak Panksepp, Thomas Fuchs, and Paolo Iacobucci, "The Basic Neuroscience of Emotional Experiences in Mammals: The Case of Subcortical FEAR Circuitry and Implications for Clinical Anxiety," *Applied Animal Behaviour Science* 129 (2011), 1–17; Jaak Panksepp, "Affective Neuroscience of the Emotional BrainMind: Evolutionary Perspectives and Implications for Understanding Depression," *Dialogues in Clinical Neuroscience* 12 (2010), 533–45.

28. Jaak Panksepp and Jules B. Panksepp, "The Seven Sins of Evolutionary Psychology," *Evolution and Cognition* 6 (2000), 108–31.

29. Joseph LeDoux, "Rethinking the Emotional Brain," *Neuron* 73 (2012), 653–76.

30. Steven Pinker, *The Language Instinct: How the Mind Creates Language* (New York: William Morrow, 1994).

31. Panksepp, "Affective Consciousness."

32. Bernard J. Baars, *A Cognitive Theory of Consciousness* (Cambridge, U.K.: Cambridge University Press, 1988).

33. L. F. Haas, "Phineas Gage and the Science of Brain Localisation," *Journal of Neurology, Neurosurgery, and Psychiatry* 71 (2001), 761.

34. Sergio Paradiso et al., "Frontal Lobe Syndrome Reassessed: Comparison of Patients with Lateral or Medial Frontal Brain Damage," *Journal of Neurology, Neurosurgery and Psychiatry* 67 (1999), 664–67.

7. The Concept of Complementarity: The Gift from Physics

1. John Tyndall, *Fragments of Science for Unscientific People: A Series of Detached Essays, Lectures, and Reviews*, vol. 1 (New York: D. Appleton and Company, 1871), 119.

2. Joseph Levine, *Purple Haze: The Puzzle of Consciousness* (Oxford: Oxford University Press, 2001), 6.

3. Ibid., 87.

4. David J. Chalmers, "Facing Up to the Problem of Consciousness," *Journal of Consciousness Studies* 2 (1995), 200–19.

5. John Tyndall, "The Belfast Address," in Tyndall, *Fragments of Science*, vol. 2 (London: Longmans, Green, and Co., 1879), http://www.gutenberg.org/files/24527/24527-h/24527-h.htm#Toc158391647.

6. William James, *The Principles of Psychology* (1890), in Hutchins, Adler, and Brockway, *Great Books of the Western World*, vol. 53, *William James*, 97.

7. Ibid., 95.

8. Ibid.

9. William Stukeley, *Memoirs of Sir Isaac Newton's Life* (manuscript, 1752; facsimile, Royal Society, 2010), retrieved June 26, 2016, from http://ttp.royalso ciety.org/ttp/ttp.html?id=1807da00-909a-4abf-b9c1-0279a08e4bf2&type =book.

10. John Conduitt, Draft account of Newton's life at Cambridge (1727–28), Keynes Ms. 130.04 (Cambridge, U.K.: King's College), retrieved June 26, 2016, from http://www.newtonproject.sussex.ac.uk/catalogue/record/THEM00167.

11. Rudolf Clausius, *The Mechanical Theory of Heat—with its Applications to the Steam-engine and to the Physical Properties of Bodies* (London: John van Voorst, 1867).

12. Max Planck, "On the Law of Distribution of Energy in the Normal Spectrum," *Annalen der Physik* 4 (1901), 553.

13. Helge Kragh, "Max Planck: The Reluctant Revolutionary," *Physics World* 13 (2000), 31.

14. L. Piazza et al., "Simultaneous Observation of the Quantization and the Interference Pattern of a Plasmonic Near-Field," *Nature Communications* 6 (2015), 6407.

15. Richard Feynman, Sir Douglas Robb Memorial Lecture (University of Auckland, 1979), retrieved September 2, 2016, from https://www.youtube .com/watch?v=xdZMXWmlp9g.

16. Richard Feynman, Messenger Lecture: "The Quantum View of Physical Nature" (Cornell University, 1964), retrieved October 3, 2016, from https://www .youtube.com/watch?v=x5RQ3QF9GGI.

17. John von Neumann, *Mathematical Foundations of Quantum Mechanics*, trans. Robert T. Beyer (Princeton, N.J.: Princeton University Press, 1955).

18. Feynman, Messenger Lecture.

19. Jim Baggott, *The Quantum Story: A History in 40 Moments* (Oxford: Oxford University Press, 2011), 100.

20. Ibid.

21. Robert Rosen, "On the Limitations of Scientific Knowledge," in John L. Casti and Anders Karlqvist, eds., *Boundaries and Barriers: On the Limits to Scientific Knowledge* (Reading, Mass.: Perseus Books, 1996), 203.

22. Ibid.

8. Non-Living to Living and Neurons to Mind

1. Howard Hunt Pattee, "Physical and Functional Conditions for Symbols, Codes, and Languages," *Biosemiotics* 1 (2008), 147–68.

2. Howard Hunt Pattee and Joanna Rączaszek-Leonardi, *Laws, Language and Life: Howard Pattee's Classic Papers on the Physics of Symbols with Contemporary Commentary* (Dordrecht, The Netherlands: Springer, 2012), 7.

3. Ibid., 8.

4. Pattee, "Physical and Functional Conditions for Symbols, Codes, and Languages."

5. Howard Hunt Pattee, "The Physical Basis of Coding and Reliability in Biological Evolution," in Pattee and Rączaszek-Leonardi, *Laws, Language and Life*, 33–54; repr. from *Towards a Theoretical Biology 1, Prolegomena*, Proceedings of an International Union of Biological Sciences symposium, Bellagio, Italy August–September 1966, ed. C. H. Waddington (Edinburgh: Edinburgh University Press, 1968), 67–93.

6. Pattee and Rączaszek-Leonardi, *Laws, Language and Life*, 10.

7. Howard Hunt Pattee, "Physical Problems of Decision-Making Constraints," in Pattee and Rączaszek-Leonardi, *Laws, Language and Life*, 70; repr. from *International Journal of Neuroscience* 3 (1972), 99–106.

8. Pattee, "Physical Basis of Coding and Reliability."

9. John von Neumann, *Theory of Self-Reproducing Automata* (Urbana: University of Illinois Press, 1966); Erwin Schrödinger, *What Is Life? The Physical Aspect of the Living Cell* (1944; repr. Cambridge, U.K.: Cambridge University Press, 2012).

10. Steve Martin, https://www.youtube.com/watch?v=q_8amMzGAx4.

11. Howard Hunt Pattee, "The Complementarity Principle in Biological and Social Structures," inPattee and Rączaszek-Leonardi, *Laws, Language and Life*, 143–54; repr. from *Journal of Social Biology Structures* 1 (1978), 191–200.

12. Howard Hunt Pattee, "Cell Psychology: An Evolutionary Approach to the Symbol-Matter Problem," in Pattee and Rączaszek-Leonardi, *Laws, Language and Life*, 170; repr. from *Cognition and Brain Theory* 5 (1982), 325–41.

13. Marcello Barbieri, "Biosemiotics: A New Understanding of Life," *Naturwissenschaften* 95 (2008), 579.

14. Ibid., 597.

15. Ibid.

16. Ibid., 580.

17. Ibid., 596.

18. Noa Liscovitch-Brauer et al., "Trade-off between Transcriptome Plasticity and Genome Evolution in Cephalopods," *Cell* 169 (2017), 191–202.

19. Christian B. Anfinsen, "Principles That Govern the Folding of Protein Chains," *Science* 181 (1973), 223–30.

20. Von Neumann, *Theory of Self-Reproducing Automata*, 77.

21. Pattee and Rączaszek-Leonardi, *Laws, Language and Life*, 13.

22. Rudolf K. Allemann and Nigel S. Scrutton, eds., *Quantum Tunneling in Enzyme Catalyzed Reactions* (Cambridge, U.K.: Royal Society of Chemistry Publishing, 2009).

23. Indranil Chakrabarty and Prashant Prashant, "Non Existence of Quantum Mechanical Self Replicating Machine," *arXiv:quant-ph/0510221v6*, 2007.

24. Niels Bohr, "The Quantum Postulate and the Recent Development of Atomic Theory," *Nature* 121 (1928), 580.

25. Pattee, "The Complementarity Principle," 144.

26. Ibid., 149.

27. Ibid., 153.

28. Feynman, Robb Memorial Lecture.

29. James, *Principles of Psychology*, in Hutchins, Adler, and Brockway, *Great Books of the Western World*, vol. 53, *William James*, 117.

30. Pattee and Rączaszek-Leonardi, *Laws, Language and Life*, 28.

31. Ibid., 11.

32. Feynman, Robb Memorial Lecture.

9. Bubbling Brooks and Personal Consciousness

1. James, *Principles of Psychology*, in Hutchins, Adler, and Brockway, *Great Books of the Western World*, vol. 53, *William James*, 149.

2. Matthew E. Roser et al., "Dissociating Processes Supporting Causal Perception and Causal Inference in the Brain," *Neuropsychology* 19 (2005), 591.

3. Sherrington, *Man on His Nature*, 275.

4. Niels Kaj Jerne, "Antibodies and Learning: Selection versus Instruction," in *The Neurosciences: A Study Program*, eds. Gardner C. Quarton, Theodore Melnechuk, and Francis O. Schmitt, pp. 200–205 (New York: Rockefeller University Press, 1967).

5. Alan M. Leslie and Stephanie Keeble, "Do Six-Month-Old Infants Perceive Causality?" *Cognition* 25 (1987), 265–88.

6. Thomas Nagel, "What Is It Like to Be a Bat?" *Philosophical Review* 83 (1974), 435–50.

7. Liane Young and Rebecca Saxe, "Innocent Intentions: A Correlation between Forgiveness for Accidental Harm and Neural Activity," *Neuropsychologia* 47 (2009), 2065–72.

8. Michael B. Miller et al., "Abnormal Moral Reasoning in Complete and Partial Callosotomy Patients," *Neuropsychologia* 48 (2010), 2215–20.

9. Neil Young documentary, part 2, retrieved August 27, 2016, from https://www.youtube.com/watch?v=Lslh6hG9EVQ.

10. Steven Pinker, *How the Mind Works* (New York: W. W. Norton, 1997), 133.

11. Jaak Panksepp and Lucy Biven, *The Archaeology of Mind: Neuroevolutionary Origins of Human Emotions* (New York: W. W. Norton, 2012).

12. Jaack Panksepp, "The Periconscious Substrates of Consciousness: Affective States and the Evolutionary Origins of the SELF," *Journal of Consciousness Studies* 5 (1998), 566–82.

13. Andrew R. Barron and Colin Klein, "What Insects Can Tell Us about the Origins of Consciousness," *PNAS* 113 (2016), 4900–8.

14. Nicholas J. Strausfeld and Frank Hirth, "Deep Homology of Arthropod Central Complex and Vertebrate Basal Ganglia," *Science* 340 (2013), 157–61.

15. Robert B. Barlow Jr. and Anthony J. Fraioli, "Inhibition in the *Limulus* Lateral Eye *In Situ*," *Journal of General Physiology* 71 (1978), 699–720.

16. Shreesh P. Mysore and Eric I. Knudsen, "A Shared Inhibitory Circuit for Both Exogenous and Endogenous Control of Stimulus Selection," *Nature Neuroscience* 16 (2013), 473–78.

17. Diane M. Beck and Sabine Kastner, "Top-down and Bottom-up Mechanisms in Biasing Competition in the Human Brain," *Vision Research* 49 (2009), 1154–65.

18. Steven D. Wiederman and David C. O'Carroll, "Selective Attention in an Insect Visual Neuron," *Current Biology* 23 (2013), 156–61.

19. Paul Buckley and F. David Peat, *Glimpsing Reality: Ideas in Physics and the Link to Biology*, rev. ed. (New York: Routledge, 2009), 134.

20. Bisiach and Luzzatti, "Unilateral Neglect of Representational Space."

21. Denise Barbut and Michael S. Gazzaniga, "Disturbances in Conceptual Space Involving Language and Speech," *Brain* 110 (1987), 1487–96.

22. Gary Taubes, *Good Calories, Bad Calories: Fats, Carbs, and the Controversial Science of Diet and Health* (New York: Anchor Books, 2007).

10. Consciousness Is an Instinct

1. Howard Hunt Pattee, "Can Life Explain Quantum Mechanics?" in Ted Bastin, ed., *Quantum Theory and Beyond: Essays and Discussions Arising from a Colloquium* (Cambridge, U.K.: Cambridge University Press, 1971), 307–19.
2. Rosen, *Life Itself*, 13.
3. Ibid., 111.
4. Ibid., 112.
5. William James, "What Is an Instinct?" *Scribner's Magazine* 1 (1887), 355.
6. Ibid., 356.
7. Joshua T. Vogelstein et al., "Discovery of Brainwide Neural-Behavioral Maps via Multiscale Unsurpervised Structure Learning," *Science* 344 (2014), 386–92.
8. James, "What Is an Instinct?," 359.

ACKNOWLEDGMENTS

One of my favorite stories is about the young aspiring opera star singing for his first time at Milan's La Scala. After his debut the audience yells, "*Ancora, Ancora.*" He smiles to himself and belts out the song again. And again, the audience yells, "*Ancora.*" This goes on for four or five encores, and finally he turns to the audience and says, "Wait, I have sung the song now five times. What more do you want?" Some guy in the balcony yells back, "You will sing it until you get it right."

When I finished my previous book—a scientific memoir of sorts that told the story of split-brain research—I thought I was finished with books. It was an enjoyable book to write because the perspective flowed from personal experience and was sprinkled with stories that remain a big part of my life. It turns out that imbedded in that book were the beginnings of another book, this book. As one reader put it to me, "So now that your personal story is out of the way, write about consciousness per se." That is a different assignment and one that requires hard work, new work, and a lot of help.

There is one person who helped in the project like no other, and that is my sister Rebecca, a part-time MD, a part-time botanist, a part-time scientific writer and researcher, and a full-time bon vivant. Everybody and everything she touches turns into something better than it was. She started working with me on my various books right after a major surgery I had in 2006, kind of helping out with the editorial aspects. Quickly, she became fascinated with neuroscience and soon enough became a research assistant as well. Her intellectual clarity, her passion to know more and more about everything, and her good cheer have become central to everything I have done ever since, and I remain in her debt.

When Bridget Queenan arrived at our university to galvanize our brain initiative, I knew the everyday humdrum of academic life was going to change for the better. Her unyielding wit, drive, and intelligence, combined with her sensational editorial skills, came in to add to the effort. There were others as well. Of course, I always try to turn my graduate seminars into exploratory settings for my projects. During the first year, students bring in new material ideas, and for this endeavor, one student in particular, Evan Layer, was extremely helpful. During the second year, the new class serves as critics, editors, and more for the developing chapters.

Over the years, one develops a small group of professional friends who will actually read various drafts of the book and will comment in detail, like true friends, with no sparing of the rod. I am in debt to Walter Sinott-Armstrong, who tried to keep me inbounds on my philosophical thoughts, to Michael Posner, Steven Hillyard, Leo Chalupa, John Doyle, Marcus Raichle, and many others, who struggled to help me keep the brain story straight. And finally, to my wife, Charlotte, who really keeps me on the straight and narrow. Her influence is everywhere.

After all the internal checks, the book goes off to New York City and the publisher. This is my first book with FSG and I hope not the last. Editors Eric Chinski and Laird Gallagher both encouraged and critiqued with clarity and force. After a first go around, I found their edit so compelling that I asked them for another round. I didn't know they were both trained philosophers and that they read my efforts with more than general interest. They read it with knowledge, and untold times helped me with clarity. And then came the copyediting by Annie Gottlieb. Every line was passed through her relentless mind for both accuracy and comprehension. I am indebted to one and all, and I should point out that so are those who choose to read my efforts.

INDEX

A NOTE ABOUT THE AUTHOR

Michael S. Gazzaniga is the director of the SAGE Center for the Study of the Mind at the University of California, Santa Barbara. He is the president of the Cognitive Neuroscience Institute, the founding director of the MacArthur Foundation's Law and Neuroscience Project, and a member of the American Academy of Arts and Sciences, the National Academy of Medicine, and the National Academy of Sciences. He is the author of many popular science books, including *Tales from Both Sides of the Brain*.